WORKING FOR LIFE
Careers in Biology
SECOND EDITION

WORKING FOR LIFE
Careers in Biology
SECOND EDITION

Thomas A. Easton

Plexus Publishing, Inc.

Contents

Preface

You are just beginning to look into your future. You are asking yourself what you will do with your life. You are wondering how you will earn your living, how you will fulfill your deepest yearnings, perhaps even how you will find fame.

You may be a college student. More likely, you are still in high school. Right now, you face a more immediate decision than the choice of a life–long career. You must decide what to study in college, or perhaps even whether to go to college at all. You are handicapped, and you know it, by not knowing all the possibilities. On the other hand, you are not inhibited by knowing too well the long, hard years of study and preparation needed to qualify for those possibilities.

You bring to your decisions only a knowledge of what your relatives and friends do, a sense of professional roles gained from TV, movies, and books, and a folklore concerning adulthood you share with your peers. You may know how incompletely and inadequately you are equipped. This book will try to help, to open a window on one aspect of the landscape of possibilities before you.

You also bring to your decisions the knowledge you have gained from your high school courses. You have studied English, history, politics, French, math, art, and science. You have participated in sports. You have debated, played chess, and sung. You have even worked part-time and summers. You have tasted a dozen or more different fields, and you have some idea of what the workers in these fields do with their lives.

But you know that before you can decide what to do with your own life, you need more information. You need to know how and where to train, how to start, how far you can go and how well you will be paid, who you might work for, and how stable the job market is. You want to know how many different careers there are in a field, and you want to know how *you* might fit in.

That is the function of *Working for Life: Careers in Biology.* It cannot hope to tell you everything about every area of human endeavor. It *does* tell you the breadth, stability, and prosperity of one field, biology, and describe certain areas within that field, such as the health professions and biological technology (e.g., genetic engineering), that promise faster growing and more lucrative career opportunities.

And this little book lives up to its promises. Reviewers called it "valuable and refreshing . . . the best on the market." Unfortunately, its statistics and salary data were soon out of date, and a second edition became essential. Yet this new edition is not just an update on the numbers. It recognizes changes in certain trends the first edition thought important, and it discusses current topics such as AIDS. It adds some information, including a new section on resumes, and rearranges more. But despite the changes, the aim remains to serve you, the reader, by bringing you the information you need to choose intelligently among the careers open to you.

1

What Is Biology?

You have some idea of what biology is. You have had one course in it, a broad overview. You may have had more. You know biology is the study of life. It is also one of the sciences, and in this sense it is like chemistry and physics, which study the nature of matter, or astronomy, which studies the stars and planets. In fact, biology is often called a subdivision of chemistry or physics, because life is but one aspect of the behavior of matter. The same can, of course, be said of astronomy, and psychology and sociology can be called subdivisions of biology.

To be sure, there may be only one, single science, the study of the universe in all its aspects, and the naming of the aspects may be artificial and arbitrary. Yet scientists have found the distinctions of the names useful. Chemists, biologists, and psychologists, for instance, do their work without confusion except at the borders between their fields. They argue jurisdiction only in such areas as biochemistry (the chemistry of life) and physiological psychology (the biological mechanisms of thought and feeling).

Is biology only the study of life? No. Biologists are by no means all researchers or scholars. Many are teachers. Many are more akin to engineers, for they are concerned with applying biological knowledge to agriculture and medicine. They seek new crops, disease treatments, ways of attacking pests, weeds, and problems of nutrient shortages. Others are managers, concerned with running large research or production programs; or salesmen, selling the products of pharmaceutical and other biology-based firms; or bureaucrats, administering grants and regulating pollution, wildlife exploitation, and hazardous research. Still others comprise the hordes of technicians who do the donkey-work of lab and field. And then there are the millions of doctors, nurses, and other health-care personnel.

Biology clearly encompasses a great many jobs and careers and occupies many thousands of people full-time. In fact, it may be the one field that offers the widest range of possible careers. Train as a biologist, and you retain the broadest range of choices for your future. Unless the whole field bores you—and it may—it is thus a good direction for you to take if you cannot make up your mind.

Biology may also offer the greatest job or career security. Biologists in general work with the most vital of human needs—food and health. There is thus a continuing and growing need for biologically educated teachers, technicians, researchers, health workers, and so on. There is even a growing need for biologically sophisticated nonbiologists, for biology affects more areas of life every year—in environmental issues, in new technologies such as genetic engineering, in the politics that issue from population pressures and food shortages—and newspapers, magazines, and TV all need writers who understand and can explain what is going on.

Demand, Supply, and Biologists

Just how great is the demand for biologists? First, let's say that the broad category "biologists" includes subcategories called life scientists (those we often think of when we think of biologists— physiologists, microbiologists, etc.), conservation workers, environmental scientists, and health workers. The last group is certainly the largest, for there were some 5.7 million people in health-related occupations in 1984, of which 156,000 were dentists, 1,377,000 were registered nurses, and the rest were physicians, licensed practical nurses, optometrists, veterinarians, and so forth. The people in all

Table 1: Distribution of life scientists by sector of employment.

	1970	1982
Business and industry	23%	33%
Educational institutions	43	33
Federal government	20	17

these groups, because they are concerned with obtaining and using knowledge of living things, are biologists.

The non-health groups are smaller. The two main groups within the conservation category are foresters, who plan and supervise forest growth and use, and range managers, who manage, improve, and protect rangeland resources to maximize their use without damaging the environment. In 1984, these two groups together numbered 25,000.

According to the Commission on Professionals in Science and Technology, in 1982 there were 298,410 life scientists, split into 210,000 biological scientists, 63,500 agricultural scientists, and 27,500 medical scientists. Employment in these fields increased 48 percent between 1970 and 1982. In the same period, biologists increased their numbers by 108 percent in business and industry, by 25 percent in educational institutions (includes teachers), and by 90 percent in the federal government. Clearly, the demand is rising, most quickly in business and industry and in the federal government.

It is worth comparing life scientists to scientists and engineers of all kinds, whose numbers went up by 50 percent between 1970 and 1982. Most of this group are employed in business and industry, with only 391,800 of the nation's 2,972,400 scientists and engineers engaged in teaching. Life scientists are now equally split between industry and education, although in 1980 there were 60 percent more life scientists in education; the difference is due largely to the rise of biological engineering firms. In industry, only about a third of all scientists and engineers are engaged in research and development; the rest work in such areas as sales and management.

3

Wildlife biologist Ray Owen holds a bald eagle. (Photo courtesy of the University of Maine.)

Table 2. Life scientist employment by area and sector of employment, 1982. (From *The Technological Marketplace*).

	Business & indus.	Educational insts.	Federal govt.	Nonprofit organizations
All scientists & engineers	2,186,400	391,800	284,400	109,800
Life scientists	110,300	109,800	57,100	21,200
Biological	71,600	75,300	45,800	14,200
Agricultural	33,800	18,200	9,200	2,300
Medical	5,000	15,800	2,100	4,600

In one area of applied biology that is not covered in these figures, the picture is even rosier. The numbers of health care personnel have grown tremendously in recent years. Between 1970 and 1984 the U.S. acquired 30 percent more M.D. physicians, 34 percent more dentists, 54 percent more veterinarians, and 84 percent more registered nurses. Nevertheless, many parts of the country are still considered "health manpower shortage areas." The growth in this area is likely to continue for the foreseeable future.

Table 3. Health personnel growth, 1970-84.

	1970	1984	
Physicians (M.D.)	334,000	435,000	(+30%)
Dentists	116,000	156,000	(+34%)
Veterinarians	25,900	40,000	(+54%)
Registered nurses	750,000	1,377,000	(+84%)

The demand for young people with biological education is high and likely to stay high. However, this may be less true in university and government research. The federal government is currently less generous with research funds than it used to be. At the same time, the population of college-age youth is shrinking as past declines in the birth rate are felt. The demand for college and university teachers in general will not pick up again until near the end of the century. However, the health professions are expanding and industry may soon need many more biologists at all levels.

Starting salaries vary considerably. In 1987, a new life sciences bachelor's graduate could get about $15,000 per year, a new master's about $23,000, and a new doctorate about $28,000 working for the federal government. They could get similar or better salaries in industry.

With experience, biologists earn more. In 1983, doctoral-level life scientists averaged $27,500 with less than five years of professional experience, $31,000 with 5-9 years of experience, $36,500 with 10-14 years, $40,400 with 15-19 years, and $42,400 with 20-24 years. Medical scientists earned still more, collecting, for instance, $39,800 with 10-14 years of experience.

Currently, new agriculture and biology graduates enjoy an average (over all employers) starting pay of $17,700 and $17,200 respectively. Research and development salaries in industry average some $11,000 per year more than teaching salaries.

Like demand, the supply of young people with biological educations is also high. In 1980, 23,000 bachelor's degrees were awarded in agriculture and natural resources and 47,111 in biological sciences. In 1982, the numbers were down to 21,000 and 42,000, respectively. In both years, many bachelor's graduates did not continue their education in their fields. In 1980 only 3,987 master's degrees and 991 doctoral degrees were awarded in agriculture and natural resources, and 6,536 master's and 3,638 doctoral degrees in biological sciences. In 1982, the score was a little better at 4,163 agricultural and 5,874 biological masters, and 1,079 agricultural and 3,743 biological doctorates.

Competition for jobs is thus high as well, but in general, science and engineering graduates at all degree levels have an easier time finding work in their specialities than social science and humanities graduates. Through the 1970s and early 1980s, the unemployment rate for men in the biological sciences stayed close to one percent; for women, it was 3-4.5 percent. For new graduates, in 1982 the unemployment rate was higher but still low, 5-10 percent for new bachelor's and 2-3 percent for new master's graduates. By two years after graduation, some 60 percent of male bachelor's, 40 percent of female bachelor's, 80 percent of male master's, and 70 percent of female master's graduates have (finally or still) jobs in or close to their fields. For agricultural, medical, and biological scientists with doctorates, in 1979, 70-80 percent had jobs in their fields, regardless of years since graduation.

Projections of numbers of job openings and bachelor's, master's, and doctor's degrees through the mid-1990s indicate that in the biological sciences, there should be enough openings to absorb all new doctoral and most new master's graduates. Many of the bachelor's graduates will have to go into health-related and other fields. Even if they do not find work related to their major fields, however, their educations will not be wasted. There is more and more demand for people with biological sophistication in all fields—in sales, engineering, management, and politics. This is not likely to change, for biology is coming to play as great a part in public life as physics has in the years since World War II. This much was predicted by Frederic Joliot-Curie, who shortly before his death in 1958 said that while the first half of this century belonged to the physicist, the second would belong to the biologist.

The History of Biology

Many people believe that history is meaningless except in terms of people. Ralph Waldo Emeson actually once said, "There is properly no History; only Biography." In this view, a field such as biology has a history only as far back as we can identify biologists. Biology thus began in the fourth century B.C., when Aristotle tried to classify living things. It continued with Aristotle's student, Theophrastus, who wrote the first botany book; with Hippocrates, a founder of the modern approach to medicine and the author of the Hippocratic Oath still taken by new physicians; and with Seneca,

Table 4. Employment (1984) and projected growth in employment per year (through the mid-1990s) for various biological occupations. (From *Occupational Handbook,* 1986-1987 edition)

Abbreviations:
a: projected growth equal to average for all occupations
a-: worse than average for all occupations
a+: better than average for all occupations

Occupation	Employed (1984)	Projected growth
Agricultural scientists	37,000	a
Biological scientists	100,000	a
Chiropractors	31,000	a+
Dental assistants	169,000	a+
Dental hygienists	76,000	a+
Dental lab techs	51,000	a
Dentists	156,000	a+
Dietitians & nutritionists	48,000	a+
Dispensing opticians	42,000	a+
EKG technicians	21,000	a
EEG technicians	5,900	a+
Emergency medical technicians	133,000	a–
Foresters and conservation scientists	25,000	a–
Health services managers	336,000	a++
Licensed practical nurses	602,000	a

Table 4. (continued)

Occupation	Employed (1984)	Projected growth
Medical assistants	128,000	a++
Medical lab workers	236,000	a–
Medical record technicians	33,000	a++
Nursing & psych- iatric aides	1,204,000	a+
Occupational therapists	25,000	a++
Optometrists	29,000	a+
Osteopathic physicians	20,000	a+
Pharmacists	151,000	a–
Physical therapists	58,000	a++
Physician assistants	25,000	a++
Physicians (MD)	435,000	a+
Podiatrists	11,000	a++
Radiologic technologists	115,000	a+
Recreational therapists	17,000	a+
Respiratory therapists	55,000	a+
Registered nurses	1,377,000	a++
Science technicians	239,000	a
Speech pathologists & audiologists	47,000	a
Surgical technicians	36,000	a
Veterinarians	40,000	a+

who in the first century A.D. used crude magnifying glasses to look at living things. It leaped the hiatus of the Dark Ages and picked up again with the Renaissance, with the anatomist Vesalius; with William Harvey and his discovery that blood circulates in the body; with Anton van Leeuwenhoek, inventor of a simple microscope and discoverer of microscopic organisms; with Hooke, Schwann, and Schleiden, inventors of the cell theory; with Pasteur, Koch, Darwin, Mendel; and with many more.

Recite the names and their accomplishments, and you do indeed recite a history of biology. But that history stops far short of a complete account of the development of this area of human knowledge. A more useful history recites ideas and attitudes, and these can be traced much further into the mists of time. Too, ideas and attitudes may be much more of a key to progress than names, although names—individual people—are certainly necessary.

The study of life surely began when primitive humans first used their intelligence to become familiar with the plants and animals in their lives. However, we can only guess what they knew before that time when people learned to write down their thoughts in words and pictures. Some of those guesses are quite good, for preliterate peoples around the world domesticated dogs, horses, sheep, goats, and various food plants, and they must have had some notion of how to go about selective breeding, even if they had no notion at all of why it worked. Certainly modern primitives display an intimate awareness of their environments rivaling that of the professional ecologist.

The earliest records of anything we might call biology are religious and mythical records. Genesis is typical when it answers the natural question of life's origins by saying it was created from inert matter. Perhaps because Genesis' authors well knew the phenomena of plant and animal reproduction, it then says at least one of life's purposes is to multiply. But the authors of Genesis did not study life in any modern sense. They simply named what they saw, drew some obvious conclusions, and then used their creative imaginations to make sense of it all. They probably knew more than they wrote, but their words are all that remain to us of their knowledge.

The words of Genesis have had profound effects. Some historians, for instance, feel that we should blame the Earth's current ecological difficulties on the Judeo-Christian heritage, with its "divine" message that humanity is a special creation with the right and even duty to lord it over all the rest of the living world. The Bible implies that the Earth and all its resources exist for human benefit. It says little about human responsibility to the Earth, and it encourages the human tendency to explosive population growth. However, to trace all our environmental ills to these philosophical roots is to oversimplify. Famines, wars, diseases, and environmental crises have marked the histories of all cultures. To err is human.

According to Genesis, life is a spark fanned into existence by the breath of God. Belief in such a "vital spark" dominated the minds of all who thought about or dealt with life until only a very few centuries ago. People began to question old beliefs with the Renaissance, and there arose a view that competed with and has now largely supplanted the old "vitalism." This view is *mechanism*. It sees life as a process, as an interaction of parts, of organs and molecules and atoms, as a property of certain physical systems. It can be summarized in the statement that all living things are extremely subtle and complex chemical machines.

Where vitalists split the living from the nonliving world and asked philosophical questions about their relationship, mechanists see a continuum and ask more pragmatic questions: How do heredity, reproduction, and respiration work? Why is malaria found in some parts of the world but not in others? What physical factors cause cancer? And so on. In the process, they have debunked the old idea that worms and mice can arise spontaneously from nonliving matter, and explained how one species can give rise to others through evolution.

It should be no surprise that the mechanist's view of life fits into a time of broad social change. It arose when people were learning that many ideas accepted for generations were wrong and that new things and concepts could be created. They were in love with mechanism, and the mechanistic view was a natural result. Evolution, the mechanists' final argument, arose during the Industrial Revolution, when a sense of change and progression in human affairs had prepared Darwin and his colleagues to recognize and accept similar change and progression in the natural world. Darwin's ideas then blessed human progress as natural, not just desirable and

11

convenient. Darwin shot down the vitalists, but he affirmed the "progress" message. He gave the Industrial Revolution the extra bit of momentum that today makes it so hard to bring under control.

History as ideas can also be illustrated with more specific and less wide-ranging examples. For instance, the modern issue of intelligence testing began with the effort to devise ways of identifying students who needed extra help. The resulting tests were then used to say some races were superior to others, not surprisingly confirming the testers' prior prejudices. With the rise of the civil rights movement and of a general sense of social justice, the problems with the early tests and the errors in their use, as well as with existing theories of intelligence, were recognized. The problems have not yet been solved, but there is a new sense of openness around the questions.

The future may see society and science coming together in a new view of the world. This new view will dwell less on progress, growth, and constant expansion than on balance. The concept of balance has roots in the Bible, where one view of the ideal life pictured "each man with his vine and fig tree," but it has been expanded upon and refined in the relatively new field of biology called ecology. It emphasizes not constant progress, but progress toward a steady state in which the human population uses no more resources than it can renew. It sees a more slow-paced life in a world of intimate interconnections, in a world that is a single interrelated system

Clearly, biology is more than the study of life. It reflects and shapes human thought and society. It both follows and causes changes in society, and a knowledge of biology is of great help in understanding these changes. It casts a unique light on past, present, and future. Yet it *is* the study of life, and as such it offers an understanding of how living things work and can be manipulated. It offers vision, but it also offers tools for the control of disease and the production of food and other goods.

The Fields of Biology

We have already said that "biology" includes both research and application, the academic study of plants, animals, parasites, physiology, etc., and the pragmatic use of knowledge in agriculture, medicine, and industry. Yet such a division has far more to do with the setting and goal of one's work than with its content. That is, a research-oriented microbiologist will spend his or her days in a lab striving to understand the lives of microorganisms. He may or may not teach, collect soil samples from forests or mountains, or do other things. His lab may be in a university, a drug company, or a government agency. The applications-oriented microbiologist may also work for university, industry, or government, or for a hospital or public health agency. He or she will study microorganisms in order to seek new antibiotics, to control or improve the production of beer or bread, to understand the causes of disease and hence to improve treatment, or to identify the unknown cause of some disease. In any case, however, he is a microbiologist. The arena of his labors is the same.

Looking at the subdivisions of biology in terms of these arenas or subject matters may give the potential biologist a more useful guided tour. Let us, then, take that tour. We cannot cover every biological specialty, but we can touch on a few. If you want more information, you might consult *Biology in Profile: A Guide to the Many Branches of Biology* (New York: Pergamon Press, 1981, $9.50), sponsored by a committee of the International Council of Scientific Unions and edited by P. N. Campbell.

Zoology is the scientific study of animals. It encompasses taxonomy or systematics, the identification, description, and classification of animals; mammalogy, the study of mammals; ichthyology, the study of fishes; herpetology, of reptiles; entomology, of insects; protozoology, of protozoa; and more. If he is interested in hormones, he is an endocrinologist; in behavior, an ethologist; in parasites, a parasitologist; in evolution, an evolutionary biologist. He or she may also be a geneticist, anatomist, or physiologist; a specialist in cellular physiology, biochemistry or molecular biology. But in every case, his primary interest is animals, their structure and function and nature.

Like the zoologist, the *botanist* can specialize in single kinds of plants or in single aspects of their biology. Some botanists focus on ferns, mosses, or trees; on taxonomy, evolution, or genetics; on physiology or growth regulation; on ecology, reproduction, or biochemistry.

Microbiology concentrates on microorganisms, on bacteria, yeasts, fungi, protozoa, and one-celled algae. The microbiologist too can be a taxonomist, geneticist, physiologist, or ecologist. Often, he is interested in applications. Microorganisms are used to produce antibiotics, foods (wine, beer, bread, tofu), and industrial chemicals. Many (though far from most) microorganisms cause disease, and some microbiologists are concerned with diagnosis, treatment, and prevention.

The *physiologist* cares most about how organisms work, how they respond to changes in their external and internal environments. Physiology thus takes all of life as its subject matter as it seeks an understanding of life's basic physical and chemical mechanisms. It wants to know what happens inside an organism, but also how and why it happens. It is thus a very useful tool for the biologist concerned with illness. Its subfields include general, mammalian, comparative, reproductive, and environmental physiology.

Ecology is concerned with the interrelationships of organisms and their environment. By its nature, it is interdisciplinary, encompassing many areas of biology, chemistry, geology, and meteorology. Once called "the subversive science" because it challenges our accustomed view of ourselves as independent of the world, it is necessary to human survival. The reason is simply that without it, we cannot hope to understand or forestall disastrous impacts of our activities on our air, water, and food supplies, and even on climate. Its message is well summarized by "There ain't no such thing as a free lunch."

The *pharmacologist* is concerned with how drugs work. He or she is thus a kind of physiologist, but one with a bent toward health applications. He is also a biochemist and often works with "pure" biochemists who use the pharmacologist's understanding of drug action to probe the workings of the cell.

Ethology is the study of animal behavior. Unlike psychology, it is more observational than experimental. Like psychology, it offers understanding of ourselves and of the evolutionary, adaptive reasons for our actions.

Immunology is the study of how the body defends itself against invasion by, for instance, disease organisms. Its greatest gift to medicine to date is vaccines, but it may have greater ones in store.

Immunologists study the production of antibodies, proteins that attach to foreign molecules, and the activities of immune-system cells that attack and kill foreign cells. From their efforts have come ways to produce pure antibodies (from "hybridomas") and stimulate or suppress the immune system. Soon, they may give us better treatments for cancer, more reliable organ transplants, and better relief from allergies.

Genetics is the study of heredity. As such, it has discovered the rules of inheritance and taught us how to breed improved crop plants and animals; it was responsible for the "Green Revolution." It has also discovered the importance of DNA and found out how to transplant genes from one organism to another (this branch of genetics is better known as *molecular biology*). "Genetic engineering" companies have sprung up in recent years to use this knowledge to produce human hormones and proteins for medical use. The future may yield crop plants that need less fertilizer, new foods, improved varieties of plants and animals, and more. Yet many geneticists remain basic researchers concerned mainly with working out how various characteristics are passed from parent to offspring. They usually study conveniently short-generation organisms such as viruses, bacteria, fruit flies, and mice. They may also look at human genetics, and then they may provide new explanations of troubling diseases such as schizophrenia.

There are many other fields of biology, but these are enough to show the great breadth of the field. If you have any inclination to science at all, you can surely find a career in biology, and one with a fair amount of variety within it. Most biologists split their time among research, teaching, administration, and writing. Some split theirs between university and industry. Others work full time for the government, an educational institution, a private research outfit, a zoo or botanical garden, a museum, or—you name it. There are biologists everywhere, and they do just about everything that connects at all to life.

Organizations of Biologists

Biologists, like most people, tend to band together with others who share their interests. We are not here concerned with social groups or hobby clubs, but with "professional associations." There is at least one for every biological specialty, and there are several that embrace many areas at once.

For physicians, there are the American Medical Association and the many state medical societies. For other kinds of biologists, there are the American Association of Anatomists, American Society for Microbiology, American Physiological Society, Society of Systematic Zoology, and many more. (You will run into some more in Chapter 3.) These associations often belong to larger organizations, such as the Federation of American Societies for Experimental Biology (FASEB) or the American Institute of Biological Sciences (AIBS). Individual biologists can join the AIBS, whose avowed goal is to enlist all professional biologists, or the American Association for the Advancement of Science (AAAS), which also enrolls scientists in other disciplines. Scientific societies often publish their own journals to serve as an outlet for their members and as an official voice for their fields. The AIBS publishes *Biosciences;* the AAAS publishes *Science.* Both journals have a multidisciplinary orientation and carry help-wanted ads that can help young—and old—biologists find jobs.

Most biological societies also have annual meetings that let their members present papers and meet each other. FASEB meets in April; the AAAS in January. Other society activities include placement services, low-cost insurance, information on education and careers, and professional directories. The AIBS is an especially good source for students to consult. FASEB, AAAS, and AIBS also support the Scientific Manpower Commission, an excellent source of employment statistics.

Many societies restrict their membership to biologists involved in reseach in the society's field. The multidisciplinary societies, such as AIBS and AAAS, do not. (FASEB's members are other societies only.) Some, such as Sigma Xi, which publishes *American Scientist*, are just as multidisciplinary as the AAAS, but still require proficiency in research as a prerequisite for membership. State or local "Academies of Science" are the least restrictive of all, often demanding no more than interest.

Many specialty societies draw their members from all over the world, despite the "American" in their names. Still, they do have a North American emphasis. On the international level, there is only one main organization, the International Council of Scientific Unions (ICSU). It serves to set standards of measurement and to coordinate international meetings where scientists from many lands can meet and exchange information.

However, you are not yet a biologist. You are still contemplating the field, and you are still years away from joining a professional society (though many *do* have student memberships). At this point in your life, the main value of these societies to you is as a source of information. Do you want to know more about a field? Write the AIBS or a more specialized society. You will find some addresses with the list of career pamphlets at the end of this book.

2

Who Are the Biologists?

If you now think you would like to become a biologist, you should ask yourself whether you are really suited to the life. If you are not, you are bound to be less happy in biology than you might be in some other field. You might even be an outstanding failure.

If you are suited to life as a biologist, you can expect your future to hold a totally unreasonable amount of satisfaction. Biologists, like other scientists, often marvel at how lucky they are to be paid for doing something they would do even as a hobby. They aren't in it for the money, but to satisfy an addiction to learning, to prying new secrets from Nature's treasure chest, even to using already known secrets to solve new problems or to spreading the word of life to students and the lay public. A few of them do seek wealth, but most biologists are motivated from within.

This is true no matter what a biologist does, although teaching and basic and applied research do vary in their potential financial rewards. Teachers rarely get rich. Neither do basic researchers, those who seek only to understand the phenomena of life. The applied researchers, on the other hand, seek ways to use this understanding. They want new products, new drugs, new production methods, and new devices, and when an invention pays off they can get rich indeed.

Many teachers do some research. Most researchers do some teaching, and basic and applied researchers—scientists and technologists, to some—share each other's concerns. The boundaries blur and become arbitrary, all the more when we realize that the observation of a new phenomenon can lead directly to a new invention. That happened when Alexander Fleming observed that the bread mold *Penicillium* secreted a substance that could kill *Staphylococcus aureum* bacteria. He himself did not invent antibiotics, but he might have. As it happened, the work of turning penicillin into a useful medication fell to others, largely to satisfy the needs of war.

The link of observation to invention is even clearer in the case of recombinant DNA technology, for here the inventors (genetic engineers) are the discoverers. The finding of "restriction" enzymes that can cut and splice DNA strands at specific points made it immediately possible to transplant genes from one organism to another and, for instance, to grow human proteins in bacteria. Many of the basic researchers who first discovered how to do this are now involved in biotechnology firms dedicated to applying the new knowledge.

The Basic Orientation

The basic researcher and the teacher share one concern with the applied researcher. They do not worry over whether some new fact can lead to an invention, but they do want their findings and teachings to help in the understanding of old observations and new questions. That is, all three kinds of biologists are in fact concerned with usefulness. This may be especially clear in the case of the teacher, who tries to shape a field to be clear and unambiguous so that the novice student can remember and understand. The teacher may set aside a new, but confusing, observation until other biologists have worked out its implications and answered at least some of the questions it raises. This has happened many times, most recently with "intervening sequences," chunks of DNA within genes that are not translated into protein and seemed at first to have no function; now it is believed they permit gene regulation and even rapid evolution.

"Usefulness" is thus two, or even three, things. It is what helps the teacher teach, the basic researcher understand, and the applied researcher invent. A more fundamental shared concern of biologists may be the scientific method. All biology, all science, begins with the *observation* of some fact. When the observation is repeated often enough to show that it is no fluke and a pattern exists, it leads to a *generalization*. When the biologist tries to explain the pattern, he or she forms a *hypothesis* and checks it with formal *experiments*. If the hypothesis holds up, it may advance to the status of a *theory*. If it does not, the biologist forms a new hypothesis and checks again. At any and all steps in the process, the biologist reports his or her ideas and findings to the rest of the scientific world in formal scientific papers.

The scientific method is sometimes called the "method of guess and test," but it must be more than that. Just any guess won't do. It must be possible to check the guess by experiment. Fleming thus could not guess that his mold simply did not "permit" bacteria to grow; that guess could not be tested; there was no way to find and measure a mold's "permission." On the other hand, he could and did guess that the mold secreted some substance toxic to bacteria. This could be tested simply by taking a sample of the mold's growth medium, without any mold, and seeing whether bacteria could grow in it. (They couldn't.)

It seems clear that if you are to become a biologist, you need the ability to work and think methodically. You must be curious, inquisitive, persistent, and enthusiastic, able to seek answers to novel questions and to try again, and yet again, when your attempted answers don't work out. Remember that when Paul Ehrlich sought a cure for syphilis, he had tried 605 drugs before he came up with the famed "606." That drug has been replaced by better ones now, but Ehrlich did begin the use of chemotherapy for disease, and new drugs today may require the screening of thousands of candidates.

The Biologist's Character

Biologists are no one kind of people. They vary as greatly as the members of any other group. But they do share a number of important characteristics. One of the most important is objectivity.

Objectivity

A biologist must have an open mind. Ideally, he or she is totally objective and has great integrity. She is able "to discard a pet hypothesis every day before breakfast," in Konrad Lorenz's words. She is able to say, as Louis Flexner once said, that "although I have great confidence in the observations, . . . the interpretations . . . badly need further work to test them." Such an attitude is rare. Certainly, nonscientists often find it disturbing and even threatening. So, in fact, do many scientists who fall short of their own ideals. This is true enough that it is often said, with some justice, that a truly new idea gains acceptance only with the deaths of the older members of the field who refuse to accept it.

Honesty

In the broad sense, "integrity" means "honesty." If you have been following the newspapers for the past few years, you are aware of several scandals that have involved scientists.

• One cancer researcher claimed he could successfully transplant skin from a black mouse to a white mouse. If he had been right, he would have had a great contribution to transplant surgery. Unfortunately, he was not. When pressed to prove his claim, he used a felt-tip pen to blacken a patch of white fur. The ink came off on his boss's fingers, and he was fired.

• Several researchers have plagiarized, stealing the words and ideas of others. One, when caught, lost his job, and his superior, who had coauthored the faulty papers, did not get the prestigious new job he was all set to move to.

• Others have invented data. Caught, one had to leave research; his superior then had to check his work to find out what parts of it could be trusted. Another's fraud invalidated a large and expensive medical study.

• A dead psychologist was discovered to have made up data on which many of our conclusions about the heritability of intelligence are based.

• An anthropology department chairman was caught making drugs in his lab. He went to jail.

• And so on. There are no more sinners among scientists than among laymen, but the natures of the crimes may differ. Scientists often are encouraged to believe they have a monopoly on truth, or if not on truth itself, at least on its sources. When feeling rushed or lazy, a few of them find it easy to tell themselves, "Let's not bother with the experiment. I *know* the answer." Or they feel their monopoly sets them above the law. Yet all this is only conceit or pride. Such people do incalculable damage to science. They weaken confidence in knowledge, for no one then knows what is true and what is not. They hamper progress, for valid theories cannot be built on faulty data or fantasy. And they destroy their own careers and reputations when they are caught.

Humaneness

To these primary requirements of the biologist's character, we can add a few lesser items. One is a willingness, a readiness, to care about the plants, animals, and people on which he or she experiments. Few biologists are really cruel or sadistic, but the general public looks at reports of research that involves pain or the appearance of pain—vivisection, experimental surgery, drug tests, studies of bullet wounds and heavy impacts with cadavers—and grows alarmed. They, and some biologists, are unaware that a stressed—pained—organism does not yield fully trustworthy data (except in a study of stress, admittedly). They insist via the government on humane standards of care and treatment. These standards are set forth in the booklet, *Guide for the Care and Use of Laboratory Animals*, prepared by experimental biologists, assembled by the National Institutes of Health, and available from the U.S. Government Printing Office. These standards must be followed by all researchers receiving NIH funding. They have done a great deal to ensure humane treatment of animals in the lab. However, a better guarantee of humaneness, one that needs no enforcement, may be a love of animals, an empathy that makes *you* ache when your animals are suffering.

Intelligence

A biologist should also have a modicum of intelligence. He needn't be a genius—it is unrealistic to set a Nobel Prize as your "career objective"—but he shouldn't be particularly slow either. Mental alertness, quickness, and agility are, in biology as in any

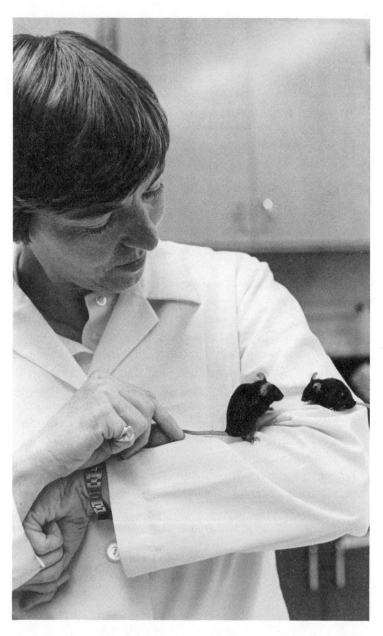

A willingness to care about and care for the animals used in laboratory experiment is a characteristic common to those who work in the life sciences. (Photo courtesy of the University of Maine.)

other field, necessary to get to the top. More specifically, a biologist should be able to reason abstractly and concretely, to solve problems and understand new facts, and to trade old notions for new. Flexibility may be more important than IQ.

We might add that advanced degrees impress many people as signs of high intelligence. Yet while some intelligence is certainly necessary to gain a Ph.D., patience and persistence are probably more essential. Since the same qualities are essential to any scientist, the degree remains a sign of qualification, a proof of aptitude as well as training.

Curiosity

Biologists must also be curious people. They must be forever asking questions, and then seeking the answers. This is clearly necessary in research, basic or applied, but it is also a very useful trait for the teacher. He or she must encourage curiosity in students, and there is no better way to do this than to show it in action in oneself.

Asking and answering can be elaborate processes, but they can also be simple. As an example, let me offer you a bit of my own experience. My specialty, insofar as I have one, is reflexes. One reflex I had read of and observed appears when cats mate. The male bites the female in the back of the neck. This automatically evokes a posture in her that facilitates the mating and puts him in an appropriate position. My question was, is this reflex necessary to the act? It certainly seemed to be involved in every feline copulation. If it were blocked, might mating be prevented? Might not interference with the reflex offer a method of contraception for housecats, one cheaper and less irreversible than spaying? The answer came when I constructed a cardboard shield to cover the back of a female cat's neck. It did in fact prevent mating. I had guessed it would, but I could not have been sure without trying it out.

This willingness to try out new ideas is the aspect of curiosity that is essential to science. It is the essence of experiment. You may find it useful to think of it in the form of a motto and an emblem: "Ask the next question" and ⊖→ The arrow in the Q means simply that you should never stop letting one question suggest another. *Always* ask the next question, for there is always one more.

Precision of Mind and Hand

Biologists must be able to detect fine distinctions, to tell closely similar plants, animals, and cells apart, and to measure things very accurately. At the same time, they must have steady hands for dissecting tiny organisms such as fruit flies, for using a micro-manipulator for cell surgery, for drawing observations, even for re-pairing equipment. However, neither of these qualities need be an in-born talent. They can be learned, and in fact this is one aim of every biology course that offers a laboratory experience.

Communication

Remember that part of the scientific method is communication of observations and conclusions to the scientific community. A biol-ogist must therefore be able to organize thoughts coherently and write them down clearly. Much of this ability comes with ex-perience, but it can also be learned in courses with names such as "Technical writing," "Advanced professional exposition," and even "Freshman composition."

Not all a biologist's communications—whether written or spoken—should be aimed at other scientists. New discoveries are often of interest to the general public and to legislators, and the bi-ologist is best off who can address these audiences without having to trust an intermediary. Journalists are well known for their misin-terpretations, oversimplifications, and premature conclusions.

Etc.

Many other qualities are helpful to the biologist. Specialties that mean a lot of outdoor work (wildlife biology, forestry, ecology, even anthropology) may call for stamina, strength, and outdoors ex-perience. Others may call for physical courage, tolerance for heat or cold, the ability to fly a spacecraft or work in isolation or in crowds. Most specialties call for the ability to work well as part of a team.

Behind all of the qualities we have discussed, however, lies a liking for living things, an enthusiasm for biology. Do you enjoy courses in the subject? Do you enjoy books and magazine articles and TV programs on it? Have you collected leaves, butterflies, or seashells? Taken part in a science fair? Done your own experiments or observations of plants or animals? If so, you may well find your professional fulfillment within the broad confines of biology.

3

The Biologist's Education

How can you become a biologist? The process is simple, but not quick. It is the same as the answer to any question of becoming: You work to acquire the skills and knowledge appropriate to your chosen specialty. And the process takes years.

We tend to think of biologists as people who have acquired Ph.D.'s. This is true enough, though it is hardly a complete picture. College and university teachers almost invariably hold doctorates, as do research leaders. But there are a great many people who can call themselves biologists who do not. Lab technicians, paramedics, and nurses may have only a two-year (Associate) or four-year (Bachelor's) college degree. High-school science teachers may have a bachelor's degree or a master's degree or the non-research oriented Doctor of Arts.

Some biologists actually have no degrees at all; they have acquired their skills and knowledge through experience alone, perhaps in the military, and they are just as well qualified as their degreed fellows. They may actually be more qualified. It is thus one of the sad facts of life that many employers insist on degrees as proof of qualifications. They ask for experience too, for that does prove fitness for a job, but they rarely settle for experience alone. Therefore, your best

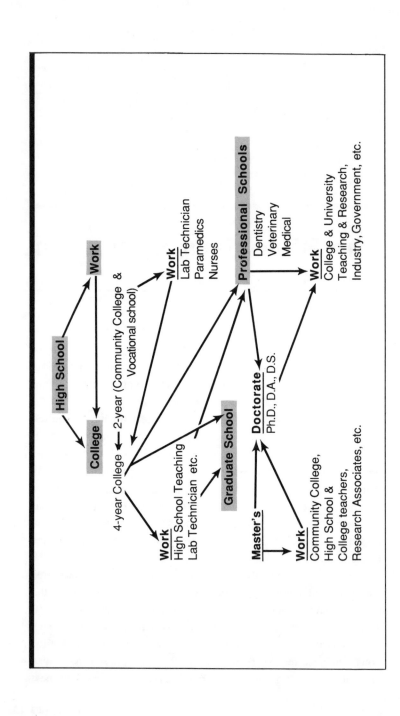

path to your chosen career will combine degree and experience. The experience may come in summer or part-time work, in assisting a professor's research, or in a few year's work between high school and college or college and graduate school. The experience will help not only in landing a job, but also in giving you a better grip on your more academic learning.

On the other hand, there are people who have *more* than a Ph.D. The research physician may hold both M.D. and Ph.D. A patent lawyer may have a law degree as well as a doctorate. A manager may add an M.B.A. to his.

The options are many, and a whole book could be devoted to analyzing them in detail. Here, in this chapter, we will focus on the traditional path to a Ph.D. We will think of the options as early stop-offs or branch points, and we will pay them relatively little attention. It may be enough to note them as in the illustration (See page 30) which shows the various pathways a high-school graduate can follow on his or her way to a career.

High School

All pathways to biological careers begin at one point, the high school. Any student thinking of making a career in biology must begin his or her preparation well before finishing this educational stage.

Certainly, as a high school student, you should take one or more courses in biology. This will provide a first real taste of the field and offer the first chance to discover whether it really appeals and what specialties are available. In addition, you should take courses in chemistry, physics, and math, up to or beyond calculus. A first course or two in computer programming will prove very useful, for there is virtually no branch of modern science that fails to use computers extensively.

Mathematics (and computers) are, according to some, one of the languages of science. Another is plain English, for the scientist must communicate clearly, concisely, and precisely. Learn to say what you mean without confusion and in as few words as possible. Unfortunately, many scientists fail on this count. Too much scientific prose is wordy, murky, turgid, jargon-laden, and confusing. It slows reading and learning, aggravates journal editors, and worsens one's chances of communicating and for promotion.

Bear in mind that everyone appreciates a good speaker and writer. Therefore, take plenty of English courses and practice your composition. In addition, read all you can, pay attention to how writers write, and try to match them. As you become a better writer, you will be pleased to find writing becomes easier and easier to do.

What about foreign languages? Most graduate schools once required a reading knowledge of two foreign tongues. Most now ask for no more than one; if a second *is* required, it may be a computer programming language such as COBOL or BASIC. The one real language should be one you will find it useful to be able to read in your field, one in which reports you wish to read are published. Once this meant German, since most non-English-language research was done by Germans. Now it means German, French, Spanish, Russian, or Japanese. The first three of these languages are offered in many high schools. The last two should be, for they are now the more useful to many scientists.

Fortunately, a great deal of foreign research is published in English, and most of the scientists you will meet at international scientific meetings speak English as a second language. However, the ability at least to read some other tongue than English will prove invaluable for more than satisfying academic requirements. You should begin acquiring this ability early on. Later, you will need the time far more for study in your specialty. Start in high school.

Languages, English, computers, math, biology, chemistry, physics. Is there anything else you should add to the list? Of course there is. *Don't* neglect your other courses, for the biologist is as much a part of the world as anyone else and must know as much about it. Still, if you do concentrate mainly on the listed subjects, you will be well prepared for further education in biology. You will even be well prepared to switch courses to another science, or to a non-science such as history or politics. Preparation for biology, because of its breadth, is preparation for almost any field you care to name. Bear that in mind if you are having trouble making up your mind. Bear in mind too that few students have chosen their careers before the junior year of college. Some are still undecided in—or after—graduate school.

You might look at the *Occupational Outlook Handbook*, or the summary leaflet *Science and Your Career*. Both come from the U.S. Department of Labor, Bureau of Labor Statistics, and survey the various careers that call for a knowledge of science. They include some non-biological careers for which biological sophistication or knowledge is helpful. The *Handbook* also covers many nonscientific careers.

Two-Year College

Upon leaving high school, many people do not go directly to a four-year college or university. They may not yet know whether they want to go at all, or they may be uncertain of their career direction. Then again, they may not be able to afford the high cost of tuition.

The high-school graduate who does not continue his or her education right away may work for a few years. This lets him gather experience, see possibilities and choose among them, and save money for later schooling. When he does go back to school, he may return tentatively, attending classes part-time or in the evening. He may very well look for a less expensive school or seek education that prepares him quickly for a better job.

The extension (or "Continuing Education") courses offered by state universities have traditionally met this student's needs. Now there are also many junior or community colleges that offer a two-year Associate of Arts degree, often with a very vocational emphasis. There are two-year programs for medical secretaries, lab technicians, beginning nurses, and many other occupations.

The great virtue of these programs is that they offer a slower, less intensive, and less expensive approach to higher education. They also prepare their students to transfer to a four-year school to obtain a bachelor's degree in two more years. They do *not* offer a lower quality education (although two-year schools do vary in quality, just as do four-year schools). They *do* offer a briefer, and hence less extensive, education. For many, they are a first step. For some, they are enough, for they do prepare one adequately for many jobs.

Biology students at four-year colleges must be willing to spend long hours in the laboratory. (Photo courtesy of the University of Maine.)

Four-Year College

Every biological specialty calls for much the same college prep-aration. This much is clear from a survey of the pamphlets put out by the various associations of biologists, each of them aimed at tell-ing students how to prepare for a particular specialty.

● The *physiologist*, who studies the internal activities of living things, needs knowledge of chemistry, biochemistry, physics, biology, physiology, mathematics, computers and communication. (*Careers in Physiology,* American Physiological Society, 9650 Rockville Pike, Bethesda, MD 20014)

● The *biological systematist* , who studies the kinds of organisms, their relationships, and their evolution, needs a broad background in all knowledge and a detailed knowledge of biology. He or she should study mathematics, computer science, chemistry, physics, geology, a foreign language, English, and all areas of biology. (*Careers in Biological Systematics,* Society of Systematic Zoology, c/o Department of Entomology, National Museum of Natural History, Smithsonian Institution, Washington, D.C. 20560.)

● The *ornithologist*, who studies birds, needs courses on birds and other organisms, math, statistics, chemistry, physics, geology, and a foreign language, and must be able to speak and write English well. (*Career Opportunities in Ornithology,* American Ornithologists' Union, National Museum of Natural History, Smithsonian Institu-tion, Washington, D.C. 20560)

● The *mammalogist*, who studies mammals from one of several viewpoints—structure, function, evolution, behavior, ecology, clas-sification, or economics—should study in college cellular and de-velopmental biology, genetics, evolution, ecology, anatomy, phys-iology, ethology, botany, math at least through calculus, chemistry, biochemistry, physics, statistics, computer science, a foreign language, and writing. (*The Science of Mammalogy*, American Society of Mammalogists, National Museum of Natural History, Smithsonian Institution, Washington, D.C. 20560)

● The *ichthyologist*, like the ornithologist or mammalogist, must concentrate on one kind of animal, in this case fishes. In addition, this specialist needs a good background in biology, physiology, embryology, comparative anatomy, evolution, behavior, ecology, genetics, organic and inorganic chemistry, math through calculus, statistics, foreign languages, and English. (*Career Opportunities for the Ichthyologist,* American Society of Ichthyologists and Herpetologists, National Museum of Natural History, Smithsonian Institution, Washington, D.C. 20560)

• The *pharmacologist,* who develops and studies drugs, needs to have studied math, physics, chemistry, biochemistry, biophysics, physiology, behavioral science, neurology, biomedical engineering, pathology, or microbiology, statistics, math, and more. (*This Is the Profession of Pharmacology,* American Society for Pharmacology and Experimental Therapeutics, 9650 Rockville Pike, Bethesda, MD 20014)

In this list, we can see occasional areas of required knowledge so specialized that most students will not encounter them before graduate school. We can also see a common requirement, for skill in communication, both written and spoken. We can see another, for computer literacy. And another, for math and statistics, and this requirement cannot be emphasized too strongly. Statistics is essential in the designing of experiments and the analysis of data. Math is equally essential to understanding and to building explanatory models or theories. In fact, the usual recommendation that students take math through calculus does not go far enough. At least one field, theoretical biology, does, and others could, make use of abstract algebras, fiber bundles, Green's functions, and other mathematical specialties.

There is a common core of knowledge all biologists need. This is the math and statistics, chemistry, computer science, zoology, botany, physiology, embryology, anatomy, genetics, microbiology, evolution, ecology, and behavior. Look at a small college's catalog, and you will find a list much like this is required of all who aspire to a bachelor's degree in biology; the list may even exhaust the courses offered by that school. Universities and larger colleges may offer many more courses; their requirements may be similar, but, with more courses available, they are likely to be more flexible, offering the student more freedom to pick and choose.

No student should feel that only those schools that are biggest, with the most varied offerings and the most glittering reputations, are worth attending. Small, relatively unknown schools can prepare you quite well indeed for a career in biology. In fact, many outstanding biologists have come from small colleges with only local reputations, and the small liberal arts schools have graduated a number of biologists all out of proportion to their collective size. The best way to judge a school may be, once you have seen that it has the essential basic courses, to ask whether its graduates have been able to do graduate work at outstanding universities.

Once enrolled in a college or university with your eye on a career in biology, you will face four years of course work. The first two years will cover a wide variety of areas. In the last two, you will specialize in your "major" field. In small schools, your major will be just biology. In larger ones, you may be able to major in zoology, or ecology, or genetics, or some other of the broad subdivisions of biology. (You will probably not be able to narrow your specialty all the way down to, say, ichthyology until graduate school.) At the same time, you will have a secondary, and less concentrated, focus on a related field, such as chemistry, math, computer science, anthropology, sociology, or psychology. This is your "minor."

Don't worry about delaying your final specialty until graduate school. Biologists need a broad view of life and its physical and social context. They *should* take the time for general courses. Only then will they be able to judge the importance of their own work and its place in the larger scheme of things. At the same time, a broad overview helps the student see the choices available and ensures—or helps—the choice of a career that will prove still satisfying years later. Remember, too, our earlier comment that biology is a field for the uncertain. A biology major emerges from a liberal arts education with many more options before her than, say, a political science or economics major. She is prepared for further, graduate education in a biological specialty, in basic or applied research, in human or animal medicine, in dentistry and agricultural science and forestry and fisheries and . . . . There is no end to the list, and it only lengthens further as the student broadens her education to include history, philosophy, economics, religion, art, music, and so forth. It is entirely possible for you to emerge from college less certain than ever of what to do in life—but far better equipped to do anything.

We should add here that every student will benefit tremendously

Table 5. Degrees in agriculture and biological sciences (National Center for Education Statistics).

	1970	1976	1982	1985	1989
		(actual)			(projected)
Bachelor's degrees					
Agriculture & natural resources	12,382	19,402	21,029	27,800	28,550
Biological sciences	37,031	54,275	41,639	56,140	55,030
Master's degrees					
Agriculture & natural resources	2,197	3,340	4,163	4,990	5,120
Biological sciences	5,800	6,582	5,874	7,170	6,570
Doctoral degrees					
Agriculture & natural resources	1,004	928	1,079	1,080	1,060
Biological sciences	3,289	3,392	3,743	3,140	2,760
Ph.D.'s as percent of bachelor's degrees					
Agriculture & natural resources	8%	5%	5%	4%	4%
Biological sciences	9%	6%	9%	6%	5%

from actual experience in his or her future specialty. This experience can take the form of serving as a teaching assistant in lab courses, often open to undergraduate seniors. It can mean helping a faculty member in research. It can mean part-time or summer jobs off campus, working in research labs, museums, parks, zoos, and so on. Such experience provides a taste of the future, and a basis for changing one's mind. It may also introduce a student to research interests that will prove absorbing throughout a long career. In addition, it can help pay the costs of a college education.

Graduate School

Graduate school is only a little more than a century old in this country. The first Ph.D. (Doctor of Philosophy) program was set up by Yale University in 1860. It is now, however, *the* mechanism for training skilled teachers, scholars, and researchers, and the Ph.D. degree marks the highest level of educational achievement.

The requirements for the Ph.D. vary some from school to school, but they include everywhere the fact of specialization. It is in graduate school that the biology major becomes an ichthyologist, geneticist, mammalogist, or anatomist. In the process, he or she satisfies foreign language requirements, assists in teaching and research, selects an original research topic, does the research, writes a dissertation (or thesis) presenting that research, and takes various examinations, including an oral defense of the thesis. The object of this ordeal is to be sure every new Ph.D. knows his or her field and is capable of original contributions to it.

Relatively few people try for their doctorates. As Table 5 shows, only 5-9 percent of those who receive bachelor's degrees go on to gain doctorates. In absolute terms, the number of bachelor's degrees has risen sharply in agriculture and natural resources since 1970, while the number of doctorates in these fields has stayed about the same. In the biological sciences, both bachelor's and doctor's degrees have grown more slowly. However, growth in all areas is expected to decline or reverse, upsetting a trend of long standing. In 1955-56, the number of biology Ph.D.'s was 1,025. In 1964-65, it was 1,928, 90 percent greater. By 1980, it was up another 90 percent, but by 1989 it will have fallen back to pre-1970 levels. However, this failure of growth *is* a projection. It may be totally wrong if the job market for biologists improves greatly, either because the economy as a whole improves or because the new biotechnology firms grow rapidly.

Not every graduate student aims for the Ph.D. Some opt for the equivalent D.Sc. (Doctor of Science). Others, planning to make a career of teaching with little or no research, choose to emphasize preparation for teaching in their studies and take the D.A. (Doctor of Arts). Many end their studies with a master's degree, which takes less time to earn—one or two years instead of three to five, or more—and requires a simpler, less original research project. Some schools treat the master's degree as a sort of consolation prize for students who decide they cannot qualify for the Ph.D. Others offer it as a legitimate end in its own right.

If you are still in high school, four years of college and five or more years of graduate school may seem a depressingly long grind to face. Yet every career requires a long apprenticeship, years of training and skill-building. This is just as true for musicians, writers, artists, businessmen, and politicians as it is for scientists. And as a scientist, you will emerge from graduate school as young as 26 or 27, or, if you have taken a few years off for work experience, in your early thirties. You will have a degree that documents your abilities. You may even, in your doctoral research, have already made a valuable contribution to the scientific literature. It is not at all unusual for graduate students to publish their original work, thereby acting as fully professional researchers.

Professional Schools

The term "vocational school" means different things to high school and college graduates. To the former, it promises training as a secretary, mechanic, clerk, lab technician, and the like. To the latter, it means "professional school"—a place to gain specialized training for a highly paid career. To the biology major, this career may be as a physician, a veterinarian, or a dentist. It is for those interested in using biological knowledge to heal, though there are also medical, veterinary, and dental basic researchers. Certainly, these professionals are biologists.

Medical, dental, and veterinary schools resemble ordinary graduate schools in many ways. They too are places of specialization, of meeting requirements, of passing tests. Where graduate school trains for research, though, the professional schools train for application. They offer exercise in the use of new learning in real situations. Medical school emphasizes experience in hospital clinics, with real patients. Dental and veterinary students get similar opportunities. Professional schools' degrees document their graduates' readiness to serve the public in their various specialties.

In the case of medicine, further specialization may follow medical school and the M.D. (Doctor of Medicine) degree. There is room for some specialization within a medical program, but neurologists, proctologists, cardiac surgeons, and the other specialists gain many of their unique skills in later internships and residencies, usually working under older experts. Even without time out for work experience, they may not be fully trained until their thirties. The youngest doctors are produced by a few schools with special programs; they accept highly qualified high school graduates and give them both premedical and medical training in six to seven years.

Footing the Bill

No education is cheap. College tuition can run over $8,000-10,000 per year, although state schools may be much less expensive. Room, board, texts, and other expenses may run your budget to over $15,000 per year. If you are lucky, your family will be able to supply all the funds you need, at least while you are an undergraduate. If you are not, you will have to find those funds yourself.

Fortunately, it *is* possible to find the necessary money. Scholarships and grants are available to those who qualify. Some are small, and some are large. Often, a college has its own list. Others come from outside sources, such as the Daughters of the American Revolution, churches, or the National Merit program. A useful source of information is always your college catalog. See also *The Student*

Guide to Federal Financial Aid Programs. This is revised yearly and can be obtained by calling 301-984-4070 or writing: Federal Student Aid Programs, P.O. Box 84, Washington, DC 20044. *Meeting College Costs* is also updated annually and is available from school guidance counselors or by writing: College Board Publicatons, Box 886, New York, NY 10101. The 1985 edition of *Need a Lift?* is available for $1 from: American Legion, Attn: Need a Lift?, P.O. Box 1050, Indianapolis, IN 46206. One more book that is available in many libraries and guidance offices is *Higher Education Opportunities for Minorities and Women.* Or you can order it from: Superintendent of Documents, U.S. Government Printing Office, Washington, DC 20402.

There are also loans. Of most interest are those with low interest rates and repayment both deferred until after graduation and guaranteed by the government. Unfortunately, these loans are limited in amount. National Direct Student Loans currently carry a 5 percent interest rate and repayment begins six months after graduation; students may borrow up to $2,000 per year. Guaranteed Student Loans permit borrowing of up to $2,500 per year at 8 percent interest. However, interest rates, loan limits, and repayment terms change from year to year. See your guidance counselor or college financial assistance office for up-to-date details.

There are loans that need no repayment at all if you work, after graduation, in a specific place or for a specific employer for a few years. For instance, medical students may pay their way by agreeing to work in out-of-the-way areas; the money may come from Washington, a state, or even, occasionally, a town that desperately needs a doctor. Many students can cover their college costs with the aid of the Reserve Officer Training Corps, though that means they must serve in the military (active duty and/or reserves) for several years after graduation.

Let's look for a moment at one specific school, Colby College, in Waterville, Maine. Colby is and always has been a high-quality liberal arts college with relatively high tuition. In the 1960s, its tuition was about $1,400 per year. For the 1986-1987 school year, its tuition was $10,430; room, board, fees, and other expenses (including textbooks, but *not* including travel) brought the yearly total for one student to $16,100.

That is a stiff bill to pay, and one that very few individual students or their families can meet out of pocket. Colby and most other schools are aware of this problem, and they can be very helpful. The first step is a "needs analysis" which takes into account the family's current income, number of children in college, expenses, and assets, and sets an "expected contribution" of the family toward the student's expenses. Since there is generally a difference between this figure and the student's expenses, the school then offers student loans, parent loans, and part-time jobs in the college bookstore, food service, administration offices, and academic departments by which the students can earn up to $1,300 per year. The school also adminsters scholarships and grants, including the federal Pell Grants and Supplemental Educational Opportunity Grants. Colby also expects to supply about three million dollars in grants from its endowment in 1986-87.

Are there other ways to find a college education? Periodically, we hear calls for an alliance between industry and academia, so that more students may study while working. Many companies already pay tuition and allow time off for studies for their employees, and the future may see this grow. If it does, a college education may take longer unless a company is content to pay for a student's education in return for a promise that the student will work for that company afterward.

Certainly, no student should ignore the possibilities of part-time work, on or off campus. It can pay your bills and give you valuable exposure to your field. Many such jobs are available to the biology student, perhaps more than to students in other areas. When they come from the college or university, they may offer a stipend *plus* a cancellation of tuition charges, but this is more common in graduate school.

You should note that many people recommend that a college freshman not try both to work and to study. College is very different from high school. You will need to adjust to a new social and physical environment, to a heavier work load, and to new study habits. Until you have finished this adjustment, you should not—*if possible*—make things harder for yourself by taking on extra work. Unfortunately, this does mean that you should have your first year's expenses in hand when you start college. This may be difficult, if not impossible, and the situation does not seem likely to improve. Costs do keep going up.

Fortunately, graduate school is less of a financial problem. If you are promising student, many schools will go to great lengths to help you complete your studies. They have teaching and research assistantships, paying both stipend and tuition. They have access to training grants, industry and foundation money, and so on, and they use these funds to keep the best students. This means that you need not worry, but it also means that much of the teaching you receive in graduate school may be from older students. Your nominal professors may rarely emerge from their labs, preferring to let their teaching assistants handle what they see as a chore. This needn't be all bad, however. Some researchers are stimulating lecturers, but some are not. Their assistants may actually be much better teachers.

The professional schools are less helpful. There are scholarships, but more often students must borrow the money they need. During 1978-79, 8,800 medical students received scholarships, while over 33,000 received loans. The total amount of money involved was some $264 million, two thirds of it from the federal government. Loans can come from banks, the Health Professions Student Loan Program, or the Department of Education's Guaranteed Student Loan Program. Scholarships can come from the schools, the Department of Defense's Armed Forces Health Professions Scholarship Program (funds to be repaid by service as a commissioned officer), or the National Health Service Corps Scholarship Program (funds to be repaid by working for at least two years in a "health manpower-shortage area"). Most medical schools discourage part-time work because medical study is so demanding.

4

The Academic World

The popular image of the biologist puts him or her in front of a college or university class, or in a lab on campus. The image fits, but it is not complete. The academic world is broader than classroom and lab, and not all biologists work in the academic world.

What is the academic world? Certainly it includes teaching and research at colleges and universities. It also includes teaching in high schools, though most high school biology teachers don't get counted in the statistics. In addition, it includes university-related hospitals, clinics, and health centers, zoos, botanical gardens, museums, and private research outfits. Most of these organizations are not run for profit, they share a role in the educational process, and they often are closely associated with universities.

As we noted in Chapter 1, in 1982 a third of life scientists worked on campus as teachers and researchers. The figure used to be larger—in 1980, it was 45 percent—and it seems likely to decline further for two reasons. One is the increasing activity in biological and medical technology as researchers learn to use their knowledge of molecular genetics and immunology to produce old and new drugs, industrial chemicals, and medical tests. The other is the slowdown in growth—and the actual shrinkage in many areas—of university and college faculties, due largely to the slowdown in the growth of student enrollment. A 20 percent decline in the number of 18- to 24-year olds began in 1980 and will last until about the year 2000. Faculties no longer need to grow just to meet rising demand for education, and in fact the Bureau of Labor Statistics projects a 10 percent decline in faculty employment by 1995.

In other words, if you want to teach or do research at a college or university, you had better be prepared for stiff competition for relatively few available jobs. You might do better to set your sights on other jobs, plenty of which are available. In 1982, 67 percent of all life scientists did *not* work for educational institutions, and this majority often made better money. Faculty pay is adequate, but it is not luxurious. In 1984-85, full professors in the life sciences averaged $35,053 per academic (nine-month) year in state colleges and universities and $37,268 in private schools. Associate professors averaged $28,749 and $28,705. Assistant professors averaged $23,705 and $23,695. Instructors averaged $19,122 and $16,300, while research technicians began at $11-15,000.

But let's take a look at the various possibilities within the academic world. They include high school and college teaching, research, zoos, botanical gardens, museums and private research.

Teaching in the High School

The high school science teacher is by no means always a biologist. He or she must often teach biology, chemistry, physics, and other courses. He may even coach one or more sports in smaller schools. Often his primary training is in education, perhaps with a focus in biology or in science.

The nation's high schools need about 3,000 new biology teachers every year to fill vacancies created when people retire or leave for other jobs or when schools expand, and the supply is not enough. One study notes that there was a severe shortage of high school science and math teachers in the mid-1980s. The situation was worst for mathematics, physics, chemistry, and computers, but biology positions were going begging too.

High school teachers must generally have a master's degree. To keep their jobs, they must continue their education with periodic summer courses. This "continuing education" ensures that they remain up-to-date in their field, a task as important for teachers as for researchers. They must also stay current during the rest of the year by reading journals and books. From the AIBS, the National Association of Biology Teachers, and the National Research Foundation, they can learn of new ideas in teaching, curriculum changes, and biological discoveries. Just as useful are magazines such as *Science News, Scientific American,* and *Biology Digest.*

To qualify for a job in a public school system, you must have studied teaching methods. Some courses in this area are available to undergraduates, but if you want to be a high school teacher you are probably best advised to concentrate on a specialty area. Take courses in biology, chemistry, physics, and math. Take a few education courses if you can, for then you will be able to get a job immediately after graduation. However, most of your education training will come in a master's program. Some schools offer five-year programs in which you can earn both a bachelor's and a master's degree. Along the way, you should gain all the experience you can in teaching. You can get some of this in courses that include practice teaching. You will also find it helpful to gain some experience in a biological job; work summers in a medical or research lab, on conservation projects, in museums or zoos, and so on.

Working on Campus

The demand for college and university biology teachers is less. Only about 1,000 new faculty in this area are added each year. The requirements are more stringent, for, except in two-year colleges, a faculty member is expected to be both teacher and researcher. He or she must therefore have a doctorate. She must also keep up with her field, but she is not expected to take "continuing education" courses. She meets this need by independent study, by reading, researching, and attending scientific meetings. She has time for all this because the four-year college teacher's course load is less than the high school teacher's. The two-year college teacher's load is intermediate,

47

largely because she is not expected to spend so much time in scholarly pursuits.

If you wish to teach on campus, you will need to follow the path outlined in Chapter 3 to obtain your doctorate. Your first job will probably carry the title of instructor, and you may hold it while still a graduate student. Later, you will rise to assistant professor, associate professor, and full professor. In the process, you will probably move from school to school around the country. It often seems that whether you work in academia, industry, or government, your present employer never recognizes your true worth. Promotions come faster when you change jobs.

As your rank rises, so will your pay. You will gain the chance to teach more advanced courses and train young graduate students. You will find yourself putting more time into institutional committee work, dealing with curriculum, student and job applicant screening, and many other matters. You will also find yourself spending time writing grant proposals to gain funds for your research. These last two items are widely regarded as the main curses afflicting faculty members.

However, it is no longer as easy as it once was to find jobs on campus. A single advertisement for an assistant professor to teach introductory biology can draw more than 650 applicants. Competition is intense, and relatively fewer openings are available. Many new life science doctorates must now find alternative jobs, at least for awhile. One of the most popular alternatives is the "postdoctoral appointment."

Many new Ph.D.'s go directly from graduate school to a "postdoc." They ask a senior researcher in their field for a position in his or her lab (or they answer ads in such journals as *Science*) where they can concentrate on research. They benefit from the senior researcher's guidance and skills, and they enjoy a position free from the demands of teaching (usually) and committee work. Later, they may move into a regular faculty position.

The National Research Council calls a postdoc "an important period of transition between formal education and a career in research" (*Postdoctoral Appointments and Disappointments*, p. 1). It is a temporary job whose primary purpose "is to provide for continued education and experience usually, though not necessarily,

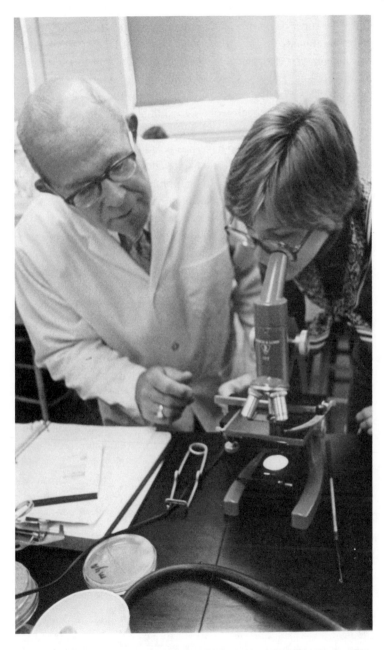

Professor Darrell Pratt guides a microbiology student. (Photo courtesy of the University of Maine.)

under the supervision of a senior mentor. Included in the definition are appointments in government and industrial laboratories which resemble in their character and objectives post-doctoral appointments in universities. Excluded are appointments in residency training programs in the health professions" (p. 11). Sir P. B. Medawar says that the postdoc is unqualifiedly a good thing, for it hones the raw, new doctorate's judgment and knowledge and builds valuable contacts. Many people who have had postdocs agree, but they also say that they were underpaid and exploited as research assistants for their "senior mentors." An additional problem is that few postdocs are women or minority-group members.

There were 6,853 life science science postdocs in 1983. In 1979 their average pay was only $12,000, less than two-thirds that of their classmates on faculties. In the same year, almost two-thirds of new biosciences Ph.D.'s planned on taking postdoc jobs, often because other, more desirable jobs were unavailable. They hoped to have better luck later on, helped by the additional training. The postdoc thus very much represents a "holding pattern" for new doctorates. As such, it has a serious cost. The unemployment rate for former post-docs is about three times that for nonpostdocs, only a third as many former postdocs gain tenure, and, perhaps because of their late start in "real" jobs, former postdocs never do make as much money as their classmates. However, though all this may make a postdoc seem a handicap to your career, it is very valuable to a future researcher. In major research universities, 73 percent of new assistant professors have held postdoctoral appointments.

Where does the money come from to pay postdocs? Often, it comes in the form of a National Cancer Institute, National Institutes of Health (NIH), or National Science Foundation (NSF) fellowship or training grant. Sometimes it comes from the senior mentor's re-search funds. However, there is such a thing lately as the unpaid postdoc. Here, the individual is expected to find a government or foundation grant to pay his or her salary, or to do without. All the "host" institution or mentor offers is lab space.

The prospects for faculty jobs will remain poor at least for the next few years. Still more new doctorates will enter the postdoc holding pattern, and many new doctorates may seek jobs outside academia entirely. Perhaps a side-effect will be an upgrading of high school biology teachers, as more doctorates enter that area, and a displacement of master's level teachers out of science completely. On the other hand, the new doctorates may well find room in industry.

Campus-Affiliated Jobs

Medical

Not all biologists employed by colleges and universities occupy traditional academic slots. Every school has a clinic or health center and hires doctors, nurses, aides, and medical technicians. Medical schools either have their own hospitals or are closely associated with public or private hospitals at which their physician-professors can practice medicine and in which medical students can learn the clinical side of medicine. These hospitals employ the full range of health-care personnel.

Museums

Some schools have or are affiliated with museums of natural history, and even unaffiliated museums are educational institutions. Sometimes faculty members also work for the museum as naturalists, taxonomists, collectors, or curators. Sometimes these people work only for the museum; then their duties do not include much teaching, although they still require a Ph.D. for upper-level positions. Their work is primarily research, but they may offer courses to the general public (including students) in museum classrooms and help local schools teach their students, either in the museum or in the schools.

One profession that exists only within museums is that of "museum exhibitor." This person conceives, plans, designs, and sets up the exhibits with which the museum speaks to the public. He or she must be a generalist well grounded in general biology, especially systematics, ecology, and conservation, and geology, geography, paleontology, anthropology, and principles of design. She must both understand the subjects she deals with and be able to work with specialists in many areas.

Although museums don't employ very many people, they do offer a pleasant, satisfying working environment. If the prospect appeals to you, you might seek a job as a technical assistant. It requires a bachelor's degree in biology, and with additional training (sometimes offered by the museum), it can lead to higher positions, even to the job of curator. For more information, write the American Association of Museums, 2306 Massachusetts Ave., NW, Washington, D.C. 20008.

Zoos, Botanical Gardens, and Arboretums

There are also zoos and botanical gardens or arboretums. "Living museums" for animals and plants respectively, they may be run by cities, states, or countries. A few private ones also exist. In all cases, they employ experts skilled in caring for and breeding plants and animals. The botanical garden hires people who collect, preserve, and organize live and dead plant specimens from all over the world, as well as researchers in ecology, systematics and economic botany, plant physiology, genetics, anatomy, and pathology, and biochemistry, among others. The zoo hires veterinarians, animal breeders, geneticists, ecologists, animal behaviorists, and so on. These people may travel all over the world in search of specimens, on research in the wild, and to visit other zoos.

Both kinds of living museums hire people with Ph.D.'s for their more specialized positions. If you have a bachelor's or master's degree, you can become a gardener, groundsman, horticulturalist, caretaker, animal keeper, exhibit preparer, publication writer or editor, or librarian. In any position, you will be involved in research and education, just as in a museum of natural history, but you will be free of many bureaucratic regulations and of campus committees. Your pay will be comparable to what you could get on campus.

Private Research Laboratories

Our last segment of the academic world lacks most of the educational mission of the other sectors, although it may offer university students part-time jobs, training programs, and even lab space for graduate research, and its staff members may also serve on faculties. This is the private research laboratory. It may, like the collection of labs at Research Triangle Park in North Carolina, be affiliated with or run by one or more universities. It may be totally independent, like the Jackson Laboratory in Bar Harbor, Maine, famed for its work in mouse genetics, cancer research, and more.

Other private labs include the Marine Biological Laboratory in Woods Hole, Massachusetts. Biologists from all over the world flock there every summer to do research and study. The pace is quieter the rest of the year, but research does continue, along with specimen collecting and other activities. The Salk Institute of Biological Studies in San Diego, California, is named for Jonas Salk, inventor of one of the first polio vaccines and the Institute's director. The Worcester Foundation for Experimental Biology in Shrewsbury, Massachusetts, is well known for its development of the birth control "Pill," but its people work in other areas of biology as well.

The biggest problem a private lab faces is funding. If it's lucky, it may have an endowment large enough to finance the lab with the interest. It may, like the Jackson Lab, have a salable product (specially bred mice). It may, like the Marine Biological Laboratory, offer summer courses and training programs. It may do research on contract for government agencies or industry. It may also live on private donations or on grants from foundations and other sources such as NIH.

Private labs hire biologists of all kinds and all levels. They need administrators, technicians, and distinguished scientists. They even have the occasional Nobel Prize winner. They offer all a pleasant, stable research environment and pay as well as most universities. They ask for the same qualifications as universities, too, meaning a doctorate in a research specialty.

5

Working for Industry

People who work in the academic world can also work for industry, part-time or as consultants. People who work for industry can hold part-time faculty apppointments. And people can move back and forth between the two, working this year on campus and next in industry, with perhaps a stint in government before or after. The difference between academia and industry may thus be less than many people would like to think. Wherever he or she works, the scientist lives the life of the mind, inquiring into the secrets of nature. The main difference is that the industrial scientist seeks answers with immediate, practical uses. The academic scientist is free to seek answers without foreseen or foreseeable applications.

Do you have a practical bent? Do you want to apply your knowledge of biology to human needs? Certainly there are places for applied researchers on campus and in museums, botanical gardens, and zoos. There are more places elsewhere. If you go into the health professions, you can work in hospitals and clinics, as a physician, nurse, lab technician, or administrator. If you go into industry, you can work in almost any kind of company you can name, you can pursue almost any specialty you wish, and you can make far better money than in the academic world.

The Upjohn Company is a large pharmaceutical corporation that

employs nearly 8,000 people in and near its Kalamazoo, Michigan headquarters, about 12,000 people nationwide, and over 22,000 people worldwide. It makes drugs for human and animal health care and agricultural and industrial chemicals, performs laboratory tests for physicians and hospitals, and provides paramedical personnel for homes, hospitals, and long-term care facilities (such as nursing homes) through its various operations. In research, it employs biologists, biochemists, microbiologists, pharmacists, pharmacologists, physicians, and medical technologists. In manufacturing, it hires pharmacists. In sales, it uses biology, premedical, and pharmacy graduates. (It also hires in nonbiological fields.) As for management, well—the Vice Chairman of its board has an M.D. and a Ph.D., its vice-presidents include four Ph.D.'s, two M.D.'s, and a Sc.D., and the vice-president and general manager of the Agricultural Division is a Doctor of Veterinary Medicine.

Upjohn spent $280 million on research and development (R&D) in 1985, up greatly from $147 million in 1980 and $93 million in 1976. Sales in 1985 were $2,008 million, so that the company spent 14 percent of its revenue on R&D (in 1976, it spent 9.0 percent on R&D). Current research focuses on cardiovascular, central nervous system, infectious, inflammatory, and metabolic disease, hair growth and skin biology, reproductive medicine, and cancer products. The company seeks drugs to protect gastrointestinal cells, affect brain function and mental performance, and control immune system responses. It is also investing heavily in biological technology. Not surprisingly, Upjohn expects its business to prosper over the next few years, and it encourages young people to inquire about careers with the company, supplying information brochures on the company, its activities, and the jobs it offers (write to: Employment, The Upjohn Company, 7000 Portage Road, Kalamazoo, MI 49001).

The pharmaceutical industry depends utterly on its researchers for new products and continued success. It therefore gives its scientists practically everything they need for equipment, materials, and technical assistance. It also allows—even urges—its people to collaborate with people in other disciplines and on campus. It lets them publish scholarly papers and books, and often lets them hold part-time positions on local campuses.

As a whole, industry performs some three-quarters of this nation's research and development. In 1985, estimates the National Science Foundation, it spent $80 billion, up 81 percent from 1980. Figures on the number of scientists and engineers involved in this R&D effort are not available at this writing for 1985 or later, but in 1983 industry employed 535,600 such people, up 20 percent from 1980. These figures cover all of industry—food, chemicals, drugs, energy, machinery, etc. The figures for areas related to biology alone are smaller but still impressive, although precise figures are not available after 1980, when industry spent $520 million on R&D for food and related products and $1,670 million for drugs and medicines. It also employed 7,500 and 21,100 R&D scientists and engineers in these respective areas. Many of these researchers were biologists of one kind or another; other biologists worked in production, sales, and management.

Other industries that hire biologists are:

- makers of agricultural chemicals, including pesticides, growth regulators, fertilizers, and dietary supplements
- cosmetics makers
- fish breeders, harvesters, and processors
- wood growers, harvesters, and processors
- textile and leather makers and users
- petroleum products companies
- public utilities
- aerospace companies
- scholarly and textbook publishers
- makers of lab equipment and supplies
- collectors, growers, and processors of biological materials for classroom and lab use
- biological testing companies
- commercial medical labs

The work of these biologists includes testing for the effects on

Table 6. Civilian employment (in thousands) in biological occupations, actual 1982 vs. projected 1995. (Bureau of Labor Statistics)

Occupation	1982	1995-low trend	1995-high trend
Life scientists	80.5	106.4 (+32.2%)	108.8 (35.2%)
Agricultural	21.7	25.7 (18.4%)	26.3 (21.2%)
Biological	51.6	71.3 (38.2%)	72.7 (40.9%)
Medical	7.2	9.4 (30.1%)	9.8 (36.1%)

health of drugs, food additives, dyes, and chemicals, studying the effects of construction and energy projects on the environment, striving to improve plant and animal growth and to limit pests, and presenting biological information to students and the public. The biologists themselves, who are technicians, researchers, and even quite distinguished scientists, devise and test new products and check the quality of production. They are food technologists, dietitians, pharmacologists, toxicologists, ecologists, geneticists, biochemists, zoologists, botanists, and many more.

Table 6 shows the Bureau of Labor Statistics' projections of changes in civilian (non-defense) biological employment between 1982 and 1995. Note that the projections for the biological sciences are more consistently high even than those for the medical area. This is because of the enthusiasm of industry for research in biological technology, as well as because of the biological tie-ins of other technologies. Yet companies that hire biologists don't hire them only for research. They put them in sales, believing that people who understand a product and its uses can sell it more effectively and can communicate customer needs more precisely to the research staff. They use them in management, too, believing that experts in a field are better able to oversee work in that field. They pay their people well too.

Early in the 1980s, one small Boston company employed a dozen people with biological training to supply pure enzymes to research labs around the country. Four of the dozen were technicians, with B.S. degrees, making $12-15,000 per year. Four were researchers with Ph.D.'s, making $27-32,000. Two were managers; one had a Ph.D., one a Sc.D. (Doctor of Science); they made $37-45,000. The last two were administrative assistants (I was not told their degrees and pay). The fields of these biologists included genetics, molecular biology, and biochemistry. Their "minority status" breakdown was interesting: The administrative assistants were both women, and one was Black; two technicians were women, and one technician, two researchers, and one manager were of Asian stock.

One small genetic engineering company employed 29 biologically trained people. Of these biologists, eight were technicians with M.S. degrees, making $16-23,000 per year; five were women and one was Asian. Thirteen were researchers with Ph.D.'s, making $26-34,000;

Molecular biologists at a genetic engineering company isolate and splice genes. By putting genes into bacteria, they can cause the bacteria to manufacture human proteins such as hormones and interferons.

two were women and one was Asian. Two were managers making $18-22,500. Six were "other" (perhaps upper management) with Ph.D., M.A., and B.A. degrees, making $30-60,000; three were women. These people's fields included microbiology, genetics, and molecular biology, and they had experience in recombinant DNA and gene cloning.

As the eighties come to an end, salaries are somewhat higher than they were when I collected the above data. Salaries in the life sciences overall rose by about 10 percent between 1981 and 1983. Since then, however, inflation has slowed; we can expect 1987 and 1988 salaries to be no more than about 10 percent above the 1983 level. Still, in 1984, R&D life scientists with bachelor's degrees and no experience were starting work at nearly $17,000 a year; with ten years of experience (defined as years since their first degree), they were making $27,000. R&D life scientists with master's degrees and ten years of experience were making $28,000, and with doctorates, they were getting $31,000.

How does industry get its biologists? It recruits them from faculty and new graduates. It may also sponsor further education for its employees. In the drug industry, many companies may pay part-time tuition in addition to full-time salary.

Yet not all industry researchers work directly for industry. They may be college or university faculty members, splitting their time between teaching and research, but doing their research with industry funding and on problems of interest to industry. Such arrangements have always existed, but today they are becoming more extensive and common than ever, as major companies such as Hoechst AG, Du Pont, Monsanto, General Foods, and Elf Technologies pump millions into university research, often in exchange for exclusive rights to potential discoveries.

Investment on campus has declined somewhat since its peak in the late seventies and early eighties, when genetic engineering was the hottest new technology on the stock market, for the technology has proved slower to pay off than the optimists had hoped. Yet the pattern of industry-university partnership remains strong, and many researchers welcome such arrangements, largely because modern reseach often needs large amounts of money, and the federal government is supplying less and less of it. The companies welcome them because they need not build and maintain their own research facilities and they can "buy" the best people on the frontiers of potentially valuable new knowledge. However, some people worry that "buy" is the word that matters. They fear that such arrangements will compromise scientists' objectivity, diminish research diversity, reduce loyalty to teaching, and because of the need to protect potential trade secrets, delay publication and hence slow the further advance of knowledge.

Sponsoring on-campus research represents a cheap tap into the new technologies springing so rapidly from cell biology, genetics, molecular biology, and genetic engineering. At the same time, though the millions involved may seem small to industry, they are wealth untold to many researchers, and they will surely spur the growth of those fields they fertilize, making them among the most quickly growing and most lucrative areas for future employment.

The last few paragraphs emphasize the research biologist's role in industry. Let me remind you that biologists also work in quality control, sales, and management. They evaluate drug actions, analyze chemical toxicities, and monitor or project environmental impacts. In addition, they may help train sales people, technical representatives, and managers. If they can write at all well, they may work as editors and writers of internal reports, press releases, and even professional papers; most large firms, and some smaller ones, hire technical writers, the best of whom have enough scientific background to understand what they must make clear to their readers.

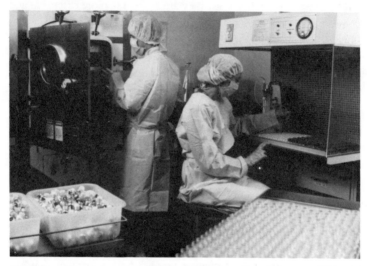

A genetic engineering company lab worker uses sterile techniques to fill bottles with genetically-engineered pharmaceutical preparation. Once in the bottles, the preparation is freeze-dried and labeled. It is then ready for use.

How do you get a job in industry? Clearly, you can develop yourself as an academic researcher in a field with commercial potential, such as genetic engineering, and let yourself be "bought" by a company eager to expand its research activities in that direction. However, this method can work for relatively few people and is to the taste of even fewer.

You would probably do better to remember some things you read back in Chapter 3. While you are in college, find part-time and summer jobs in your field. If industry appeals to you, seek them there. Perhaps you can find an employer who will pay your tuition in return for a promise that you will not change jobs for a certain number of years after graduation. Perhaps you can find a full-time employer who will encourage and help you to continue your education on a part-time basis.

The same thing applies when you are in graduate school. At the end, when you have the highest degree you want for the time being and are ready to begin your career, you will either have a job or be well acquainted with one or more possible employers and their work. You will at the very least have an "in" that can get you your first full-time job. If at this point you have only a bachelor's degree, you will be able to become a technician, manager, salesperson, or technical writer or editor. With a master's, you will find the same jobs open; you will also be qualified to be a research assistant. With a Ph.D., you will be free to become a researcher. If you gain your education while working for one employer, you may well progress from technician to research assistant to researcher, perhaps with spells along the way as a salesperson or writer. By the end of your career, you may even become a corporate vice-president or president.

Working for industry has definite advantages. Pay is a big one. So is adequate equipment, materials, and technical support. It also has disadvantages, of which the biggest may be lack of freedom. Your work is defined by the company's goals. That is, your mission is to increase sales and profits by coming up with new products or by convincing customers (and government) that your company's efforts are worthy. Clearly, you will be happiest in industry if you want the same things your company wants. If you do not, you will have to suppress your own wishes to some extent, and you may be happier in the academic world or in government.

6

Working In and Around Government

Do you think the academic world and industry are the whole show? I doubt it. There is a third sector of this nation's economy that is hard to miss: government. The federal government, in 1984, employed 2,941,000 civilians. The states employed 3,899,000, and local governments employed 9,594,000. All levels of government together employed 16,434,000 of this country's 105 million employed civilians, or 15.7 percent of the labor force.

This "public sector" employs a great many scientists and engineers. The federal government employed 307,100 of these people in 1984, of which 38,700 were life scientists. State and local governments, and nonprofit institutions, employed similar numbers of scientists and engineers, and of life scientists.

The federal government is the largest single employer of people with biological training. It uses them in virtually every area of its operations. The list below is *not* exhaustive:

- Centers for Disease Control (CDC)
- Central Intelligence Agency (CIA)
- Department of Agriculture
- Department of Commerce (National Bureau of Standards,

National Oceanic and Atmospheric Administration)
- Department of Defense
- Department of Energy
- Department of the Interior (National Park Service, Bureau of Land Management)
- Department of Labor (Bureau of Labor Statistics, Occupational Safety and Health Administration)
- Department of Transportation
- Environmental Protection Agency (EPA)
- Food and Drug Administration (FDA)
- National Aeronautics and Space Administration (NASA)
- National Institutes of Health (NIH)
- National Science Foundation (NSF)
- Smithsonian Institution
- Veterans Administration (VA)

The list of positions for federal biologists is as long as that for academic or industrial biologists. The U.S. Office of Personnel Management gives them, in its Announcement No. 421, as:

- Agricultural management specialist
- Agronomist
- Botanist
- Consumer safety officer
- Ecologist
- Entomologist
- Fishery biologist
- Forester
- General biologist
- Geneticist
- Home economist
- Horticulturalist
- Husbander
- Microbiologist
- Pharmacologist
- Physiologist
- Plant pathologist
- Plant protection and quarantine officer
- Range conservationist
- Soil conservationist

- Soil scientist
- Toxicologist
- Wildlife biologist
- Wildlife refuge manager
- Zoologist

Table 7. 1984 government employees by activity (in thousands). (U.S. Bureau of Census)

	Federal	State	Local	
Defense	1,069			
Postal service	709			
Education	15	1,707	5,314	
Teachers		(513)	(3,388)	
Highways	4	248	293	
Health & hospitals	261	676	716	
Public welfare	12	179	228	
Police	65	80	604	
Fire			316	
Sanitation		1	220	
Parks & recreation		31	219	
Natural resources	245	158	37	
Financial administ.	115	126	187	
All other	446	693	1,460	
Totals	2,941	3,899	9,594	16,434

To this list we can add editors and writers, health care personnel in military and Veterans Administration services, aides, technicians, and more. Government offers plenty of job variety, and in the biology area, that variety has grown tremendously in this century. The National Institutes of Health began in 1887 when physician Joseph J. Kinyoun received a $300 federal grant to study the cholera bacterium in an attic lab at the Marine Hospital on Staten Island. By 1901, Dr. Kinyoun was in Washington, DC, with a $35,000 budget for constructing a new lab building. In 1930, his operation became the first National Institute of Health. After World War II, the NIH took its modern form as a source of grants to outside, mostly academic researchers and an employer of its own researchers. Its budget for the 1987 fiscal year is over $6 billion, its twelve Institutes spread across a 300-acre "campus" in Bethesda, Maryland, and it funds over 20,000 grants.

Like NIH, the Departments of Agriculture, Defense, and Commerce also do a great deal of basic and applied research related to their missions. The Smithsonian runs museums and publishes a magazine. The CIA needs biologists who can interpret foreign activities in the life science areas. NSF oversees research and works to improve education in the sciences. The Department of the Interior runs the national parks and manages public lands. The Department of Labor's OSHA is concerned with occupational medicine. The EPA regulates the environmental impacts of industry's (and government's) activities; it is largely concerned with the toxicity of chemicals and pollutants. The FDA cares about the toxicity of drugs, cosmetics, and food additives and about drug effectiveness; it too is a regulator. The VA runs hospitals for veterans of this country's military and researches their particular occupational health problems. It would be impossible to describe all the federal government's biological operations in detail. It may, however, be useful to say what goes on in one of them: the Centers for Disease Control in Atlanta, Georgia. Its mission is very clearly one of public service.

The Centers for Disease Control

The CDC employs many biological scientists. Its labs and offices contain physicians, epidemiologists, microbiologists, virologists, physiologists, technicians, and more. Together, their role is to keep track of disease outbreaks, identify their causative organisms, and recommend ways to control them. They also study degenerative ills such as heart disease and cancer. Recent success stories include the part the CDC played in the World Health Organization's effort to eliminate smallpox throughout the world. Today, the only surviving samples of smallpox virus are to be found in sophisticated laboratories such as the CDC's, where every effort is made to prevent the

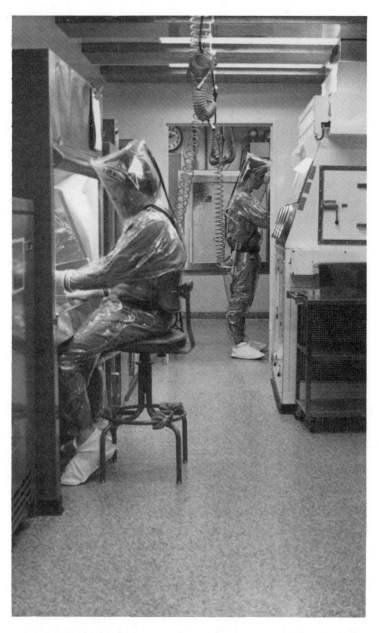

The Maximum Containment Laboratory at the Centers for Disease Control in Atlanta features isolation suits for the biologists working there. (Photo courtesy of the CDC, Atlanta, Georgia.)

virus's escape. Another success story is Legionnaire's disease, a form of pneumonia now known, thanks to CDC work, to be caused by a bacterium that grows in such places as the stagnant water of air conditioner cooling towers. Analyses of preserved tissue samples have shown that the disease is not new. It existed at least as long ago as the 1940s, though it was then thought to be just another pneumonia. It can be treated with the antibiotic erythromycin.

Every week the CDC receives thousands of specimens, many of them blood and tissue samples from disease victims, for investigation. Pathogenic microorganisms are isolated from these specimens and grown in the most stringent isolation conditions humanly possible. Tissue samples and pathogen cultures are sterilized and processed for examination by light and electron microscopy and for biochemical analyses. They are tested with antibodies to determine the past prevalence of diseases. Bacterial strains are identified, in part, in terms of their susceptibility to bacterial viruses. Computers plot the spread and extent of epidemics. And more. The purpose of it all, however, is summed up in the CDC's name—the Centers for Disease Control. All that its researchers, technicians, and administrators do is directed toward the single end of better understanding disease, its causes, effects, and spread, in order to control it and to reduce human suffering.

The War Against AIDS

In the past few years, acquired immune deficiency syndrome —AIDS—has become the one disease of most concern to many researchers and citizens, and hence to government agencies such as the CDC and NIH as well. AIDS is caused by the human immunodeficiency virus (HIV), which incapacitates certain cells of the immune system, impairing the body's ability to fight off infections and cancers and leading to death within about two years of diagnosis. Yet even though over two million Americans are infected by the virus, which is transmitted by sexual intercourse, shared drug injection apparatus, and transfusions (and by products such as clotting factors derived from whole blood), the experts are projecting that only about 250,000 people will develop AIDS between 1987 and 1992. It seems that some people are able to resist the virus's effects, at least for awhile.

AIDS is certainly a plague, and cases and deaths seem likely to continue to increase every year until after the turn of the century. But there is hope in individual resistance, in the actions of various drugs on the virus, and in various techniques for prolonging or easing the lives of its victims. There is thus a great deal of research to be done by the biologists of the CDC and the NIH. University researchers are involved as well, but the vast bulk of AIDS research is a government activity, to the tune of $300 million for fiscal year 1987.

Those who have an eye on biological and medical research positions with the federal government may take heart from a report from the U.S. Surgeon General that calls for increasing the AIDS research budget to $1 billion a year or more by the early 1990s. The National Academy of Sciences and the Institute of Medicine peg the necessary budget at more like $2 billion a year by that time. Either way, there will be plenty of jobs in this area.

The GS Grade System

Government jobs are given GS (General Schedule) or other (GM, EC, EP, etc.) "grades" according to their difficulty and responsibility. Biological jobs have grades from GS-5 to GS-15. When you get a government job, you receive the job's grade and salary. Salaries go up with grade, but if you qualify by education and experience for a GS-9 job, say, but hold a GS-5 job, you receive the GS-5 salary.

The GS-5 grade is for white-collar workers with experience or education equivalent to a bachelor's degree. The GS-7 grade requires a bachelor's degree with a B average, a class rank in the upper third of your class, or elected membership in a national honorary society (such as Phi Beta Kappa), or one year of graduate education, or one year of professional experience, or one year of preprofessional student trainee experience. The GS-9 grade requires two years of professional experience or graduate education, or a master's degree. The GS-11 grade calls for at least three years of professional experience or graduate education, or a doctorate. The GS-12 grade requires at least three years of professional experience, including at

71

least one at a level of difficulty comparable to the GS-11 level. Higher grades depend on experience and excellence. The "super grades"—GS-16, 17, and 18—are for supervisory personnel such as agency directors. You can get further information from your nearest Federal Job Information Center; check your phone book, or see your local State Job Service (or State Employment Security) Office, for the address.

Table 8. Federal white-collar (GS) pay scales and employment. (U.S. Office of Personnel Management)

Grade	Employment, 1983	Pay range, 1983	Pay range, 1987
General Schedule (GS) and equivalent	1,484,000		
Grades 1-6	541,000	$8,676-19,374	$9,619-21,480
Grades 7-10	360,000	$16,559-29,003	$18,358-32,148
Grades 11-12	362,000	$24,508-38,185	$27,172-42,341
Grades 13-15	220,000	$34,930-63,115	$38,727-69,976
Grades 16-18	1,000	$56,945-78,184	$63,135-86,682

Government employees may continue to receive full-time pay while continuing their education part-time (or full-time for up to one year). When they have finished the additional schooling, they are qualified for a job at a higher grade.

Job security, under the protection of the civil service system, is good, and there is ample opportunity for advancement. A government biologist can spend his or her entire career in the lab, field, or clinic, or move on to management positions. He is also free to publish his work in academic journals and attend scientific meetings.

The biggest drawback to government service has to do with politics. Biologists have been fired or given unsatisfying assignments for publicly disagreeing with their superiors, who may be committed to supporting a particular stance on controversial issues such as abortion or regulation of pollutants. Government employees also are restrained from helping their favorite candidates campaign for office.

Congressional Fellows

It is possible to work for the federal government and yet to be outside the traditional system. For instance, Congress deals constantly with issues of science and technology, and there are places on the staffs of every Senator and Representative and of many congressional committees for people with scientific training and knowledge. Such people may be hired through the normal channels. They may also get their jobs in other ways, as by knowing a Congressperson and helping out in campaigns or as an unpaid science advisor.

One important way for young Ph.D.'s to enter the congressional system is through the Congressional Fellows Program run by the American Association for the Advancement of Science (AAAS). The Fellows are sponsored for one year by individual scientific societies such as the AAAS itself, the American Society for Microbiology, the Biophysical Society, the American Society for Photobiology, and the Federation of American Societies for Experimental Biology (FASEB). The Program began in 1973 with seven Fellows. By 1981, it had grown to 34 Fellows, sponsored by 20 organizations and selected on the basis of excellence and interest in the legislative process. The Program's aim is to provide trustworthy technical information and advice to decision-makers.

Once chosen, the Fellows go to Washington for a two-week orientation program. They are introduced to Congresspersons, congressional staffs, Executives Branch officials and representatives of nongovernment agencies, lobbyists, and special interest groups. This is followed by a year-long seminar series.

After the orientation, the Fellows must find their own slots. Past Fellows have worked on legislation concerning health policy, environmental regulation, and energy. Once their year is done, they may return to campus or industry. They may also—and some have —remain in Washington to continue their work. Their introduction to Washington may even lead them to fairly high positions as heads of government agencies.

Lobbyists

It is entirely possible to work in Washington (or in state capitols) without being on a government payroll at all. Most companies and industries—including universities and the medical profession—have an interest in government actions, in legislation, regulation, and even policy formation. They try to affect these actions to maximize their advantage, or to minimize their disadvantage, by talking to legislators and bureaucrats. They provide information and opinion, they cajole and persuade—in short, they try to sell their own interests. They lobby, and their representatives, who may be individuals or firms, are lobbyists.

Certainly, many issues affected by government action relate to biology. On these issues, the best lobbyists must have some biological sophistication. They may even be practicing researchers or physicians, lobbying part-time either on their own or as representatives of their employers. They may also be biologists who have left their fields to lobby full-time. They are all salespeople, and their function is precisely that of a company's actual sales force: to explain technical matters understandably and to persuade the "customer" that the company deserves support and consideration.

We might note that an effective lobbyist is worth his or her weight in gold to industry. She can prevent a product ban, get a new product approved for sale, reduce the need to spend money on pollution-control or safety equipment, and so on. She can mean millions to her employer, and she can be paid very well indeed.

States and Cities

Many federal jobs are duplicated on the state and municipal level. Large cities and states run parks, zoos, museums, and hospitals. States have their own fish and wildlife, health, and environmental agencies. And they need suitable people. Unfortunately, they do not pay quite as well and offer less generous fringe benefits such as sick leave, educational aid, vacations, insurance, and retirement pay. The people who work for states and cities are often people who wish to live in a particular part of the country and are willing to trade income and other benefits for lifestyle.

Environmental Careers

Government careers clearly differ in several ways, in activity, field, and level of government. Yet there is one category of career that cuts across all the differences. Federal, state, and local governments all hire people to be concerned with the environment. These people may be researchers, technicians, writers, editors, regulators, or inspectors. They may be ecologists, systematists, foresters, biochemists, physicians, or other specialists. Their shared concern, the environment, deserves a few words of its own because it is of such broad current interest, and because environmental problems affect us all. We need clean air and water, we need to conserve natural resources, and we need to use our resources wisely.

The principal environmental scientists are human ecologists, conservationists, and planners. Each can be subdivided as follows.

Human ecologists

- Environmental health scientist—studies, analyzes, and regulates pollutants.
- Environmental engineer—designs, builds, and operates sewage and water treatment plants, controls disease-carrying pests.
- Sanitarian—inspects food handlers for cleanliness, monitors pollution, develops and manages programs to control contamination of food by pests.
- Industrial hygienist—strives to reduce workplace factors (pollution, radiation, noise, insects, fungi) that adversely affect human health.
- Health physicist—monitors and regulates public exposure to radiation from x-ray machines, nuclear reactors, etc.
- Public health physician—focuses on health problems related to poor sanitation, work conditions, and environmental factors.

Conservationists

- Agricultural engineer—develops machinery, equipment, and methods to improve the efficiency and economy of food production, processing, and distribution.
- Fisheries conservationist—studies fish biology and maintains fish populations by stocking and regulation.
- Forester—studies and manages forest resources.
- Range manager—protects, develops, and regulates the use of grazing lands.
- Soil conservationist—studies soil biology, chemistry, and physics and helps farmers (and others) protect their soil against deterioration and erosion.
- Wildlife conservationist—studies wildlife biology and maintains wildlife populations by stocking, culling, and regulation.

Planners

- Landscape architect—plans, develops, and manages outdoor areas.
- Land use planner—develops criteria and programs to control how land is used; the urban planner focuses on future city growth and development.

Most of the above job titles can fit at several levels. For instance, with a bachelor's degree, an industrial hygienist might work as a work-site inspector of OSHA; with a Ph.D. or M.D., he or she might be a researcher at NIH or a VA hospital. A Ph.D. sanitarian might study flour beetles for the U.S. Department of Agriculture; a B.A. sanitarian might inspect meat packers for the U.S.D.A. or for a state or city health agency.

There are also generalists, with expertise in several areas of biology and in other fields as well. They may administer programs or communicate to the public. They may be public relations workers, also known as "media specialists." They may be journalists, photographers, illustrators, technical writers or editors, or librarians. Government, like industry and the academic world, hires them all.

Government biologists, whatever their jobs, differ from biologists in industry or the academic world in their mission. It may not be unfair to say that they are people inclined to serve the general public welfare, although job security and pay may also be factors. If their careers are "environmental," that is only because they have focused their skills on that area. They could as easily focus them on other areas. However, they might not then feel quite so essential to the public welfare. They may thus be the most inclined of biologists to public service.

7

The Self-Employed
Biologist

All biologists have to pay their bills. Like everyone else, they
must buy food, shelter, clothing, transportation, medical care, enter-
tainment, and so on. They must also pay taxes.

However, they need not earn their money by working for other
people. They can be self-employed. Those who choose to pursue
careers of this sort enjoy an unusual degree of independence. They
can often live where they please, keep no one's schedule but their
own, and pursue their own interests rather than someone else's. They
can avoid academic committee work and paper-grading, bureaucratic
wheel-spinning and form-filling, political restraints on what they can
say, and some of any employee's less than glamorous chores (although self-employment has its own chores).

If this prospect appeals to you, bear in mind that self-employed
people suffer from certain anxieties and extra expenses. They have no
dependably regular pay check. They depend on selling their services
or products, and if their customers disappear for awhile or are slow in
paying, their bank accounts can get downright scrawny. The problem
is known as "poor cash flow." In addition, they must often maintain their own offices, labs, or workrooms, buy equipment and office
supplies, keep elaborate records, and hire lawyers and accountants.
Furthermore, they must part with a larger share of their income as a

Self-Employment Tax (FICA, or Social Security), since they have no employer to kick in a matching sum, and pay for their own medical insurance.

If self-employment still appeals to you, then by all means do consider it. The benefits may very well outweigh the drawbacks in your mind. Many self-employed people find their situation very satisfying. I do, though I also often wish the drawbacks were fewer.

How does a biologist work for himself or herself? Remember that for the purposes of this book we have been considering health-care workers as biologists. The answer to the question then becomes obvious in part. Physicians and veterinarians are very often—even usually—self-employed. So are some nurses, who hire themselves out to tend convalescents or long-term invalids in their homes.

Sometimes a physician or medical laboratory technician may set up a private laboratory to do medical tests for local doctors and hospitals. Such an operation can grow to hire several technicians to do the tests, while the founder becomes a businessperson and manager.

Nonmedical biologists can also set up and run testing labs, though not necessarily medical ones. The biological testing industry evaluates chemical and drug toxicities and drug effectiveness for pharmaceutical, chemical, and other companies. Some testing labs are large; some are small. Many were started by biologists who wanted to run their own businesses. However, such businesses can require large amounts of money to set up. Medical testing labs are much more within the means of most individuals.

If you want to set up a medical testing lab, you will need expertise and experience in biochemistry, microbiology, hematology (blood biology), and pathology. Given that and professional competence, integrity, and energy, you can expect adequate to great success and financial reward.

If you want to set up a lab to evaluate chemical and drug effects, you will need the same skills, plus skills in animal care and statisitcs and the ability to cope with a maze of regulations and criteria. Given that, plus competence, integrity, and energy, you too can expect success.

Other businesses can also be run by individuals. Zoologists can run pet stores and private zoos and aquariums. Botanists can operate greenhouses and grow everything from expensive orchids to ordinary house and garden plants. Plant pathologists can work as consultants to farmers, foresters, and horticulturalists, as can entomologists. Botanists and zoologists can collect or grow, preserve, and sell flowers, leaves, twigs, frogs, fish, sea urchins, and many other organisms for use in school biology labs and in research. Turtox and Carolina Biological Supply are large "biological supply houses." There is room for more, either generalized, like these two, or specialized, focusing on, say, protozoa, or eggs, or marine animals. If they grow their own specimens, they may have many acres of streams, ponds, woods, and fields for the purpose. Such operations can be one-man shows, but they can also be carried out by large corporations. Often enough, the large companies began when one biologist decided to work for himself or herself and use his skills to make an independent living.

One kind of nonmedical biologist may be self-employed more often than any other. This is the forester. He or she may work for a forest products company, a national, state, or city park service, or a zoo or arboretum. Often, however, he sets himself up as a "consulting forester" or "forestry consultant." He then serves small landowners who cannot afford to hire a full-time forester in much the same way he would a forest products company. He "cruises" the forest, inventories the trees present, assesses their size and condition, and recommends certain cultural practices (thinning, pruning, fertilizing) or harvesting schedules. His aim is to help the landowner maximize income from the land.

The self-employed forester may also raise Christmas trees or nursery stock or grow pulpwood or lumber on a tree farm. Often, he may combine such activities with consulting in order to generate an adequate income.

Occasionally, a biologist will discover that she has a talent for writing (or illustrating). She may then become a self-employed ("free-lance") writer (or illustrator) of magazine articles, textbooks, children's books, or popular science books. If she has a taste for fiction, she may turn to writing science fiction stories, and it is quite true that several members of the Science Fiction Writers of America, Inc., have biological backgrounds. However, free-lance writing is a chancy business. One survey found that most of this country's free-lance writers make less than $4,000 a year from their writing.

Writers must generally have full or part-time jobs to pay the bills. Very few earn a decent wage with their typewriters or word-processors alone. A very select few indeed earn a very good wage. Isaac Asimov, biochemist, is a superlative example. So is Michael (*Andromeda Strain*) Crichton, a novelist and screenwriter who used his skill with words to pay his way through medical school.

Biologists can also become free-lance editors, usually of textbooks. To do this, however, you must first spend a few years working for a publisher to build up skills, contacts, and credibility. The only substitute is a record as a successful writer, and if you have that, you have little reason to edit for a living. Writing is far more satisfying. Editing is then more useful as a way to smooth out the cash flow.

No matter what your brand of self-employment, you will probably find some kind of part-time job, working for someone else, useful. It will give you a more reliable, predictable income. If you are a writer, who habitually works alone, it will give you necessary and satisfying contact with other people.

Part-time employment can mean teaching. Many colleges and universities hire part-timers to handle extension or continuing education courses, or to handle those courses for which the full-time faculty is too small. For example, I teach biology, but I have also taught technical writing for an English Department that lacks full-time faculty enough to handle all the students who wish to take the course. The department therefore hires outsiders such as professional writers to carry the load. It pays them less than it does its full-time faculty, and it provides no fringe benefits, such as medical insurance, so that it saves money by not adding a suitable number of full-timers to its staff. However, it does do the part-timers, some of whom are self-employed writers, a favor by giving them a chance at some extra income.

It is not only writers who teach part-time. Occasionally, an academic biologist comes up with a discovery that has commercial potential. Rather than sell the right to exploit this discovery to an existing company, he sets up his own firm, either with his own money or with "venture capital" from banks or other investors. He may quit his academic post. More often, he will take a leave of absence, thus protecting his position and rank, while he gets his business going. He may reduce his teaching and research load. He may continue his full-time job while developing his business on the side. If he does quit completely, he may still continue to teach a few courses on a part-time basis, thus improving his cash flow and keeping his teaching skills sharp.

The most recent commercial biological discoveries have been in the area of genetic engineering. Many of the discoverers have set up or joined businesses, while retaining some connection to their home campuses. Very few have quit completely, for they do enjoy the academic environment, teaching, and research, and they are conscious of the risks inherent in setting up any business, especially one in a highly competitive field that may not prove as profitable as wished.

Part-time employment can also mean other things than teaching. The self-employed biologist can work part-time as a researcher or technician in a lab, as a pollution or sanitation inspector, or as anything else a biologist can be. However, it is worth bearing in mind that no self-employed biologist ever sees her part-time job as her career. It is a stop-gap, a temporary expedient she will discard when her self-employment becomes more successful. That day may actually come, and it is not a dream to discard lightly. But self-employment *is* risky. It may never become successful, and it may even fail.

The wisest course may be to train yourself for a full-time career working for others. If you then decide to develop your own business, do it on the side. Do not drop your job until your business is already successful, and then think twice about doing it. It may be in your best interest to minimize your risk.

Another course is to put all your energy into your business. You will then greatly increase your chances of success. If you do fail, you will have the consolation of knowing you gave it your best shot. However, you will be left with no income until you have found another full or part-time job. If you gave up high rank or position, you may find yourself having to start over near the bottom of the ladder.

Should you work for yourself or for others? The choice is yours. It depends on whether you work best on your own or with others, on whether you would rather handle the paperwork and pay for fringe benefits yourself or let others do it for you, and on whether you are comfortable with risk.

8

Finding a Job

You now have an idea of the many things biologists can do in their careers, how well they are paid, and where they can work. You may now know that you wish to be a biologist of one kind or another. You may even be wondering how to go about finding a job.

It may be a little early yet to worry about finding your particular professional niche. After all, you still have the most important part of your education ahead of you. That is, you have yet to become a biologist. But if you have been paying attention to the advice in this book, you are already thinking of finding part-time and summer jobs to help pay your college tuition and give you experience. And if we now look briefly at the ways biologists find jobs, that may help you now and prepare you for a later day.

Contacts

Both part-time and summer jobs and full-time, career jobs can be found in the same ways. The most important ways may involve personal contacts. As a student, you may approach your professors about possible teaching and research assistantships. Your department or school financial aid office may post a list of available part-time and summer positions you can apply for. In either case, you get the job because you are in the right place, close to the right people, at the right time. As you near graduation, you will find that your faculty advisor, your department, or the professor you work for knows of fellowships, postdoctoral positions, and teaching and research jobs, in the academic world, in government, or in industry. He or she may know of these positions through flyers mailed out by employers looking for new graduates or because colleagues have asked, "Do you know someone good for this job?" If your professor says, "Yes," and names you, you then have what may be the best possible credential for the job—a personal recommendation from someone the employer knows and trusts.

Even as a student, it is to your advantage to attend scientific meetings. You will meet many people there and establish your own contacts. You will find that many of these people have their ears open for job opportunities, while others are looking for potential employees. The purpose of these meetings is supposedly the exchange of scientific information, but some say that job-hunting is an equally, or more, important activity.

Later on in your career, perhaps when you have begun to establish your reputation in research and are known to other biologists, you will find that prospective employers come to you at meetings or by phone or mail. They may know you only by reputation. They may have been given your name by one of your past professors or a mutual friend. Or they may be people you have met or worked with in the past. In all such cases, the potential job comes to you because of contacts you have made.

"Contacts" jobs may be the best ones. In general, such a job means that someone who knows both you and the job has thought that the two fit each other. They are often right. When you seek a job by answering an ad or in some other "noncontacts" way, you have at best only a vague idea of whether you and the job will get along. You should therefore cultivate all the contacts you can. Work with your professors, attend meetings, meet people, collaborate when your research permits it. Avoid working alone if you can, for that insulates you from contact and may limit your choice of jobs.

Forestry students learn to use computerized mapping techniques. (Photo courtesy of the University of Maine.)

Some procedures in the lab require biologists to work under sterile conditions. (Photo courtesy of Merck Sharp & Dohme.)

Inquiries

Whenever you seek a job, you will write letters of inquiry. These letters will describe you and your background. They will also state what job you want, as specifically as possible, and say what good you can do the employer.

If your professor or a colleague has suggested you approach a certain company, say so. Write to the name given you, and make it clear that the personal contact exists. Make clear also that you know something about the company's operations and have some idea of how you might fit it.

For example, let's say that your advisor, Professor Edna Jones, has served Ajax Chemical Company's Industrial Biochemistry Division (IBD) as a consultant. She tells you that Ajax is planning to expand its efforts in genetic engineering and may well be able to use you and your brand new master's in genetic engineering technology. She then gives you the name of the IBD Director, Iosip Spaciba.

You write directly to Iosip Spaciba, not to Ajax's personnel department. You say something along the lines of:

"Dr. Jones, who has served you as a consultant, tells me you plan to expand your genetic engineering program.

"In June, I will receive an M.S. in genetic engineering technology. Dr. Jones says this degree represents precisely the training you need for your expansion. Therefore, I wish to apply for a position with you. I am especially interested in implementing new techniques of producing industrial chemicals and would like to work with Dr. Blanken, in charge of this part of your operation.

"I have worked with Dr. Jones. . . ."

Continue with your relevant experience and research projects. Enclose your resume and letters of recommendations. Say when you will be available for an interview. And have confidence that such a letter will be far more effective than one that fails to relate you to both employer and job.

Resumes and Dress Codes

The resume you send out with your letter of inquiry is a crucial document. I could easily devote an entire chapter to it, but you can find excellent discussions of resume-writing in technical writing courses and in careers-related magazines. I recommend the College Placement Council's *CPC Annual* and the *How to Get a Job Guide* published annually by *Business Week's Guide to Careers*. Both are available in your school library or placement office.

For our purposes here, it is enough to say that a resume is really no more than a formal, concise way of presenting your qualifications for a job. It lists your education, academic honors, languages, professional memberships (such as Sigma Xi and AAAS), publications, references, and job experience. This last item may be the most important, for it says that you have not just done well in class; you also know how to use what you have learned in a productive, responsible way. If none of your past jobs used your training as a biologist, they still show that you know how to be punctual, follow instructions, and get the job done. It is thus very important to have at least part-time work experience to show. It reassures prospective employers that you won't abscond to Brazil with their money.

The secret to writing an effective resume is known to engineering students as the KISS principle: Keep It Simple, Stupid! There are three basic kinds of resumes, the chronological resume, the functional resume, and the targeted resume. In each case, a good resume is only one page long, typed on white 8-1/2" x 11" paper and, though it lists your background without elaborate detail—just the facts; no essays, apologies, or excuses—it makes you look as good as possible.

A resume is thus a suit of clothes for your personal history. Following the resume patterns prescribed by books and magazines is following a dress code for your qualifications, and it works, just as does showing up for your job interview in the best clothes you own or can borrow. Slobs—on paper or in person—have a tendency to turn off potential employers.

Campus Placement Offices

Many campuses have placement offices. Sometimes they list

mainly jobs on campus open to students; such offices are helpful, but not to students who seek post-graduation jobs. More often, campus placement offices list full-time jobs in industry and on other campuses as well. They will also help you design your resume. They are then a very useful resource.

Placement offices, especially those at larger schools, also arrange for prospective employers in industry and government to send "recruiters" to campus. The recruiters arrive all on the same day or days and interview interested students who are near graduation. When they find students whose backgrounds, educations, talents, and interests match vacant niches in their organizations, they may invite these students to visit their organizations for further interviews. Occasionally, they will make job offers on the spot. Fortunate students (mostly those in engineering, these days) may have their pick of 30 or more job offers.

Professional Societies

Professional societies such as the Federation of American Societies for Experimental Biology (FASEB) and the American Institute of Biological Sciences (AIBS) also run placement services (available to both members and nonmembers). Both job seekers and employee seekers register with these services, and lists of each are made available annually.

Both the AIBS and the FASEB services also arrange candidate-employer interviews at their annual meetings. Every year hundreds of novice biologists meet prospective employers in this way, and the employers are academic, industrial, and governmental. For more information, write to: FASEB Placement Service, 9650 Rockville Pike, Bethesda, MD 20014, and AIBS, 1401 Wilson Blvd., Arlington, VA 22209.

Some professional societies publish journals that carry lists of "positions open." The AIBS publishes *Bioscience*. The American Association for the Advancement of Science (AAAS) publishes *Science*. Sigma Xi's *American Scientist* carries occasional job ads. There are also the *Bulletin of the American Association of University Professors* and the *Chronicle of Higher Education*, though they concentrate on academic posts.

The job ads cover all specialties, employers, and levels of experience. There are ads for postdocs, fellowships, instructors, assistant, associate, and full professors, research assistants and researchers, lab and agency directors, and so on.

If you are seeking a job through ads, don't neglect the larger newspapers and the weekend editions of smaller ones. They too advertise a wealth of positions. The difference—and it may be important to you if you wish to settle in a particular locale—is that many of their ads are for jobs in the newspaper's city, state, or region.

The Federal System

The federal government advertises some of its open positions in journals and newspapers and through contacts, but most of its positions are filled through the U.S. Office of Personnel Management (OPM). The process is both simple and attractive, for it allows a job-seeker to apply for many jobs at once. (Most states have similar systems).

The first step in the process is to call or write a Federal Job Information Center. Say how much education and experience you have, the kind of work you want, the locale you prefer, the lowest salary you will accept, and the dates of your military service, if any (veterans are chosen first). You will then be told whether applications are now being accepted for any jobs that might suit you. If they are, you must then fill out a Personal Qualifications Statement (SF 171), a Card Form 5001-BC, a College Transcript or OPM Form 1170/17, a Work Location Preference Form (OPM Form 1311), and, if applicable, a Supplemental Qualifications Statement Life Sciences Positions GS-9/12 (OPM Form 1170/24), a thesis description, and a list of publications or reports. You will then mail the completed forms to a U.S. Office of Personnel Management Area Office (see Table 9).

On receipt of your application, the OPM will evaluate you. If you are interested in jobs that are filled on the basis of competitive performance on written tests, you will be asked to take the tests. Your score then determines your place on an OPM "referral list." If your jobs do not require tests, you will be rated according to the

Table 9. Federal personnel addresses.

U.S. Office of Personnel Management Area Offices:

1220 S.W. 3rd St. Portland, OR 97204	1845 Sherman St. Denver, CO 80203
P.O. Box 25069 Raleigh, NC 27611	Attn: SSH P.O. Box 52 Washington, D.C. 20044

Prince Kuhio Federal Building
300 Ala Moana Blvd.
P.O. Box 50028
Honolulu, HI 96850

Regional offices exist too, in Atlanta, Boston, Chicago,
Dallas, New York, Philadelphia, San Francisco,
Seattle, and St. Louis.

U.S. Department of Agriculture offices:

Farmers Home Administration
Special Examining Unit
Personnel Division
14th and Independence Ave., SW
Washington, D.C. 20250

Soil Conservation Service
Special Examining Unit
10000 Aerospace Rd.
Lanham, MD 20801

Science and Education Administration
Special Examining Unit
6505 Belcrest Rd.
Hyattsville, MD 20782

qualifications you describe on the forms, and your name will be added to the civil service list. You will remain on the list for one year unless you update your application every 10-12 months.

Later, when some government agency has an appropriate vacancy, the OPM will give that agency a list of qualified applicants, including you. You may then be invited for an interview and receive a job offer. Your GS grade and pay will depend upon the job more than upon your qualifications (see Chapter 6).

Not all federal jobs are filled in this way. The Postal Service, FBI, and CIA have their own procedures. For jobs that exist in only one agency, you should apply directly to the agency. For jobs related to the U.S. Department of Agriculture's operations (agricultural management specialist, range or soil conservationist, soil scientist, home economist, agricultural research), you should write the appropriate USDA office (See Table 9).

The Rewards of a Biological Career

Once you have embarked on your career in biology, you will find that you have an entirely adequate income, good opportunity for advancement and recognition by your peers, and the sense of self-worth that comes from knowing you are doing useful, challenging work.

More to the point, however, you will be unlike your fellows who work at "just another job" and can hardly wait for weekends, vacations, and retirement to do what they really enjoy. You will be *doing* what you enjoy most. You will find that your career is more than a career. It is a way of life that gives you your most valued pleasures, achievements, and gratifications. As a teacher, you will be committed to educating young people, both in and out of the classroom. As a researcher, you will be constantly thinking of problems and solutions; they will even enter your dinner-table and party conversations. In either case, your work will never be far from your mind, and you will never willingly drop it. You will not retire easily. This is even true for the technician and administrator. No biologist need ever find her work boring, unless she is in the wrong line of business.

9

Careers in Action

You now know what biology is, what it takes to be a biologist, what sort of an education you need to become a biologist, where you might work as a biologist, and how well you can expect to be paid. However, you may well be at something of a loss when you try to picture what your future career might be like. Don't be surprised at this. It's only natural, for there are a multitude of choices ahead of you, and each one will shape your future.

Would it be some comfort if I told you that everything works out for the best—at least in hindsight? That is, once you have reached a certain stage of life or career, it is often difficult to look back and say how things might have worked out better. The biggest reservation behind this comment is that it does require that you be fairly happy with yourself. If you're not, things *never* work out right.

It might be more help to show you a few samples of actual biological careers. We have therefore listed a few biographies in the back of this book, including a collection of descriptions of women's careers in science. In this chapter, we wish to sketch the careers of four men important to the recent history of biology. They are Thomas Hunt Morgan, Oswald Theodore Avery, Erwin Chargaff, and Hans Selye.

Thomas Hunt Morgan, 1866-1945

Morgan was the experimental genius responsible for the concept of the gene as the unit of heredity. Long before anyone knew how DNA works, he placed the gene on the chromosome and worked out much of its behavior. His accomplishments earned him a 1933 Nobel Prize.

His career as a biologist began in childhood. By the age of ten, he had filled two attic rooms in his Lexington, Kentucky, home with stuffed birds, birds' eggs, butterflies, and fossils. He had also tried to dissect a sleeping cat (it refused to cooperate). At the State College (later the University) of Kentucky, he studied natural history. In graduate school at Johns Hopkins University in Baltimore, Maryland, he turned to experimental biology, specifically embryology. In 1890, he earned his doctorate at Woods Hole Marine Biological Laboratory on Cape Cod. The next year, he became an associate professor at Bryn Mawr, where he taught and continued his embryological research.

In 1904, Morgan moved to Columbia University in New York, where he turned to genetics and established the fruit fly (*Drosophila*) as the geneticist's favorite research subject. By 1910, he had described sex linkage. Soon thereafter, he developed the idea of linkage groups, tied them to chromosomes, and discovered crossing over, the exchange of pieces between members of a chromosome pair.

While at Columbia, Morgan developed a close-knit research team. He took the whole team to Woods Hole every summer, to Stanford University when he moved there in 1920, and to the California Institute of Technology in 1928. At CalTech, he developed a new kind of multidisciplinary, pure research biology department, but the administrative effort hampered his research until his retirement in 1942.

Oswald Theodore Avery, 1877-1955

An intensely private man who never married and socialized very little, Avery is known almost exclusively from public records and the reminiscences of friends. His work was virtually all his life, and it led to his discovery that DNA extracted from one strain of bacteria could change a second strain into the first; this helped clinch the case for DNA as the genetic material.

Avery was born in Halifax, Nova Scotia, to an English Baptist minister and his wife. When he was five, his mother fell ill and was declared dead. Her revival two hours later, with the report that "Jesus waved me back," may have impelled Avery into medicine later on.

At nineteen, Avery entered Colgate University, where he studied public speaking and the humanities. Only in 1900, when he entered Columbia University's College of Physicians and Surgeons as a medical student, did he begin his scientific education, and even then his grades were worst in exactly the areas in which he would later become famous, bacteriology and pathology.

He practiced medicine from 1904 to 1907, when he joined America's first private research lab, the Hoagland Laboratory in Brooklyn. There he learned reseach technique and earned a reputation as an outstanding teacher. In 1913, he moved to the Rockefeller Institute to work on the pneumonia bacteria that would later earn his fame. There he stayed for the rest of his career, except for vacations on the coast of Maine, where he sailed, collected ferns and wild-flowers, and painted watercolors. He retired in 1948 and moved to Nashville, Tennessee, to be near his brother.

Avery died before the Nobel Committee in Stockholm recognized the true significance of his work. Yet though he missed a Nobel Prize, he has been honored by a "noble prize" in the memories of his colleagues. He was, they tell us, very good at picking able co-workers and gently guiding them. He was a meticulous, skilled ex-perimenter. And he was unusually honest, for he never let his name go on a paper unless he had actually helped perform the experiments.

Erwin Chargaff, 1905-

Chargaff himself claims that despite his hundreds of papers and his discovery of a basic feature of DNA chemistry (base pairing), he is an accidental biologist, the feckless victim of fate, a straw blown in the wind of what he calls "our inhuman century."

Born in Austria, at the age of nine he moved to Vienna. There he began his education, concentrating on the humanities until at eighteen he began studying chemistry at the University of Vienna. When he graduated in 1928, he took the first job he could find, at Yale University, researching bacterial lipids. Homesick, he returned to Europe in 1930, settling in Berlin. By 1933, however, he saw that "the Black Plague had assumed the government of Germany" and moved to the Institut Pasteur in Paris. By 1935, he was back in America, this time at Columbia University, where he stayed until his retirement in 1975.

Once at Columbia, Chargaff became a productive researcher, although he retained a fondness for the pre-World War II style of science (few researchers, poorly funded and lightly administered, in the business for the love of it) and periodically shifted his research interests in order to keep his involvement fresh. The work for which he is best known began in the mid-1940s, when he read Avery's paper on bacterial transformation with DNA. He saw the implications immediately and shifted his own research to DNA. Not long after, his work helped James Watson and Francis Crick work out the structure of DNA. He thus feels some responsibility for the subsequent explosion of interest, results, and consequences in genetics, molecular biology, and genetic engineering.

Hans Selye, 1907-1982

Born in Vienna and raised in an area that is now part of Czechoslovakia, Selye eventually settled in Montreal, at McGill University, and deveoped the modern concept of stress. His father was a Hungarian surgeon who ran a private clinic. His mother was a strong-willed intellectual. Their influence, "the disdain for mediocrity of any sort, the admiration for anything that was outstanding in science, art or literature," was what turned young Selye toward science.

Selye attended a local religious college, where he excelled in philosophy and languages though not in other courses. His academic record improved at the University of Prague Medical School, where "the professors managed to present Nature as she was, not as described in texts." Immediately first in his class, he soon had his M.D.; not many years later, he also had a Ph.D. in chemistry. While in medical school, he performed his first scientific experiments (in his family's basement) and came up with a brilliantly "stupid" idea, which his professors thought a waste of time but which nevertheless became his life's work: he wanted to study the common factor shared by all patients, their appearance and feeling of just being sick. At McGill, he developed this idea and discovered what he called "biological stress," the nonspecific response of the body to any challenge, damaging or not. Eventually, other scientists did recognize his ideas as valuable. The idea that stress exists is now well accepted. So is the idea that stress may most severely affect one organ or organ system weakened by heredity, disease, or environment to cause "stress diseases" such as high blood pressure, heart attacks, strokes, ulcers, and even cancer, thanks to one man who followed through on an unconventional insight with dogged persistence.

Selye never really retired. He remained busy with research, teaching, writing, and lecturing until near the end of his life. He did, however, turn more to philosophy than science, though his philosophizing remained closely linked to what he learned about stress and how he came to believe people should live if they are to stay healthy.

Your career probably will not resemble any of the four above. After all, Morgan, Avery, Chargaff, and Selye lived and worked in a different age of the world. But you too will progress from early influences and interests through education to career and achievement. You may very well switch fields along the way, perhaps even more than once. You will change jobs and home locales. You will teach, research, administer, write, and more.

Would you like to see a career that might come a little closer to your own? I became interested in biology in high school, thanks to an inspiring teacher, a biologist father, and a nurse mother. At Colby College, in Waterville, Maine, I majored in biology; I also took many courses in math and chemistry. At the University of Chicago, I studied in the Committee on Mathematical Biology, which changed its name to the Department of Theoretical Biology before I earned my doctorate. While a graduate student, I began to write. After graduation, I got a job as a textbook editor while continuing to write science fiction and popular science articles. Eventually, I quit my job to write a biology text and moved home to

Maine. I became the book columnist for a science fiction magazine and a contributing editor to *Biology Digest*. I started teaching part-time, offering courses in biology, science fiction, and technical writing on local campuses. I also wrote more books, including this one, and began to do a little computer programming.

What does my future hold? It's impossible to say. I may well continue as a self-employed writer and part-time teacher for the rest of my life. I may become a faculty member somewhere, or enter industry as a technical writer, or biologist, or manager. I may go to work for a museum, or for government, or . . . . The options are wide open. This is one reason why I can never regret having become a biologist.

10

Women and Minorities in Biology

Are you a woman? A Black, Hispanic, American Indian, or other minority member? Then you should be concerned about your professional future. You will meet obstacles that never arise for white males. You will earn less money and take longer to gain rank and status. You may actually be treated as an inferior person, even though you—and anyone else with a grain of sense—know better.

This is true even in the broad fields of biology, traditionally among the most open of areas. However, this situation is improving, and you should not let fear of opposition keep you out of biology. Jobs *are* available, unemployment is low, the pay is adequate, and the situation should continue improving for years to come, for there are many people committed to improving it.

Until fairly recently, most professions in this country were dominated by white men. Their dominance was supported—and not just by men or whites—by the belief that women, Blacks, Hispanics, Native Americans, and other "minorities" (*women* are really a majority) are inferior to men and to whites. This belief had some research support, for a number of white, male scientists had studied the question. We know now that the studies were inherently biased, for they often looked at everyone as if they shared the same background of educational exposure, culture, and encouragement, which just wasn't so. Worse yet, people applied the "inferior" judgment to *all* women and minority members, forgetting that at best it was true only on the average. There was no recognition that even if a group is on the average inferior, it must still have exceptional members, and there was no attempt to give the exceptions the recognition and support they clearly deserved.

We may be able to understand this unfairness if we remember that even scientists seek to support their own preconceptions and find it difficult to escape them. The nature of the preconceptions about women stands out in such titles as *The Physiological Feeble-Mindedness of Women*, by Paul Moebius, published about 1900.

In the 1960s, many people recognized the unfairness of prejudice and set out to give women and minorities a fair chance. Unfortunately, to the minds of many, they did overreact, sometimes insisting not only that these groups were not inferior, but that they had no inferior members. As a result, some employers were forced by law or pressure groups to hire people unqualified for their jobs.

This stance was just as silly as its predecessor. Fortunately, it now seems largely to have disappeared. Anti-prejudice or "equal opportunity" efforts now concentrate on ensuring equal access to education and equal consideration in the hiring process, though there remain requirements that each group be represented in proportion to its relative numbers in the population among the employees of any outfit that receives federal money.

The saddest thing about prejudice—whether it is directed toward women, Blacks, or redheads—is that it forces upon us such a waste. Women comprise a bit more than half the population, and hence half the potential biologists. Blacks represent a tenth of the population, and, therefore, a tenth of the potential biologists. By blocking them from education and employment in biology, or in any other science, we forego a great deal of potential progress. By paying them less than they deserve, we impair their motivation and productivity, and again we lose what they might contribute to us all.

And it just isn't so that these groups are less able than white males, even on the average. They may not make it through the educational pipeline in the same numbers, but they are just as energetic, intelligent, dedicated, and hardworking as white males, and they should be recognized as such. If we can do that, knowledge will accumulate all the faster and more problems will be solved sooner. If we do not—well, we are now threatened by cancer, overpopulation, climate change, food and energy shortages, and we need solutions immediately. Even minor delays may prove disastrous.

Women

There are a great many opportunities for women in biology. In fact, if we count health workers as biologists—and they certainly must have an extensive knowledge of biology if they are to function effectively—it may be that most biologists *are* women. This is so because almost all of the 2 million registered and licensed practical nurses are women, women are heavily represented in many other health fields, and in 1983 they amounted to a quarter of all working scientists. In recent years, more men have become nurses and more women have entered other biological fields that require more training and knowledge. Women have increased their numbers on academic faculties, in research, and in medicine. They have also increased their pay.

However, though great progress has been made, we must ask whether women are yet treated equally with men. The answer is no. In an editorial in *Science* (October 9, 1981), Shirley Malcolm of the AAAS's Office of Opportunities in Science (OOS) noted that although women made tremendous gains during the 1970s minorities did not, and the battles for equal access, advancement, status, and pay are not finished.

Both the National Research Council and the National Science Foundation have studied the earnings of scientists and engineers and concluded that women consistently earn less than equally qualified men. The NSF reports that in 1984 women scientists and engineers of all levels of experience averaged over $11,000 per year less than their male counterparts. Women with less than five years of experience made only $4,000 less than the men. Women with doctorates made about $8,800 less than men.

The National Research Council's 1983 statistics show that in three fields—mathematics, computer science, and earth and environmental science—women with 5-year doctorates actually earn more

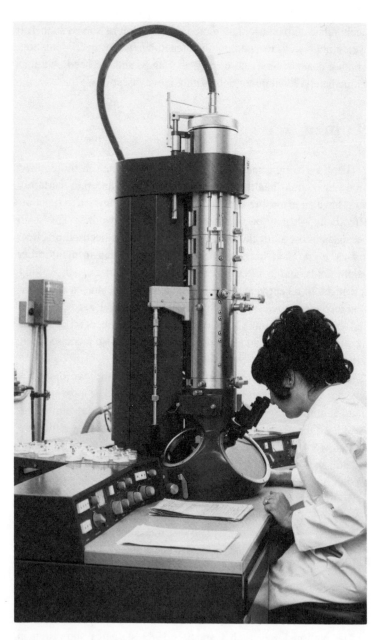

The electron microscope is a basic tool of biological research. Here a technician studies pathological changes in tissues of animals exposed to pesticides. (Photo courtesy of the CDC.)

than their male counterparts. But in general, the NRC found that the
score is the other way around. In physics and astronomy, women
with 5-year doctorates earn $2,500 less; in chemistry, they earn
$1,600 less; in psychology, they earn $2,500 less. In the agricultural sciences, the disparity is $1,200; in the biological sciences,
it is $1,900; and in the medical sciences, it is a whopping $3,000
per year.

The NSF reports that doctoral-level women earn less than men in
all fields, with the disparity being $5,200 per year in mathematics,
$8,400 in the physical sciences, $8,000 in the life sciences, and
$5,700 in psychology. The data are not consistent, though the
differences between the NSF and the NRC reports are probably
due more to the different methodologies of the surveys than to their
dates. Nevertheless, the pattern is clear: Women do not receive equal
pay for equal work.

In addition, fewer women gain tenure on university and college
facilities. This discrepancy is worst in the biosciences, where it
seems to follow from the common perception that women scientists
are less productive in research, as measured by publications, than
men. Research by Jonathan Cole and Harriet Zuckerman has shown
quite clearly that though this difference is real, it is *not* due to the
time demands of being a wife and mother: women scientists who
marry and have children publish just as much as do single women.

According to Cole, the lower productivity, rank, and pay of

Table 10. Ph.D.'s awarded to women in science and
engineering and in the life sciences,
1970-84. (National Science Foundation)

(Women/total)

	1970	1975	1980	1984
All science & engineering fields	1,626 17,743	2,836 18,358	3,801 17,199	4,568 18,069
Life sciences	538 4,165	849 4,402	1,150 4,715	1,346 4,869

women on campus may be due less to active discrimination than to exclusion from the men-only informal activities of science—"The heated discussion and debates in the laboratory, inclusion in the inner core or the invisible colleges, full participation in the social networks where scientists air ideas and generate new ones. These relationships . . . are the close collaborative relationships that . . . help to shape scientific taste and sharpen the eye for a good research problem." A similar point emerged from a 1981 OOS survey of 52 minority women scientists. Among other things, they said that "The tendency to ignore minority women when considering persons for collaboration, training or job opportunities or for quasi-social functions which enhance professional life is a problem."

This exclusion may better explain the observations that women scientists publish less than men than does the fact that they become involved in marriage and children. It therefore also explains their low rank and pay to some extent, for in science rank and pay are strongly affected by research productivity. No other difference exists to explain this. Women in science have slightly *higher* IQ's than men, they get into graduate schools in the same proportion as they apply, they have equal shots at fellowships, they get equally good jobs, and they are equally likely to be honored for their work. Only 1.4 percent of Nobel Prize winners have been women, and only 2.6 percent of National Academy of Science members were women in 1980, but this may be due to *past* prejudice and to the fact that women comprise a distinct minority of scientists.

This minority status has been the biggest obstacle for women in science. In this country, women could not even attend college until 1837. By 1875, there were only four women Ph.D.'s in the world. The situation improved greatly over the next century, but women did remain distinctly underrepresented in science. In 1965, only 744 science and engineering doctorates (7.1 percent of the total) went to women; 263 (10.4 percent) of those were in the life sciences. By 1984, the numbers were 4,568 (25.3 percent) and 1,346 (27.6 percent) respectively, a great improvement. Only the social sciences gave more doctorates to women (2,414 in 1984), although all other fields improved their performance as well. The best performing field is engineering, which gave ten times as many doctorates to women in 1984 as in 1970; unfortunately, the jump was from a minuscule 15 to an insignificant 152.

Women form more equal portions of the pools of bachelor's and master's graduates. Between 1970 and 1980, their share went from 43 to 49 percent of all bachelor's degrees, and from 40 to 49.5 percent of all master's degrees. In the same period, their shares in science and engineering went from 26.1 to 36.9 percent for bachelor's and from 17.5 to 23.7 percent for master's degrees. In 1983, women earned 38 percent of all science and engineering bachelor's degrees and 29 percent of the master's degrees. Again, the best-performing field is engineering, in which women earned 338 bachelor's and 190 master's degrees in 1970 and 9,719 and 1,900 respectively in 1983. The life sciences tally, up 124 percent over the same period, seems a trifle by comparison.

As more women have earned degrees, more women have also begun working in the sciences. However, their progress in the scientific job market has not quite matched their progress in education. In 1983, women comprised 25.6 percent of all working scientists and engineers. This does not compare favorably with the fact

Table 11. Job distribution for scientists at 4-year colleges and universities, 1983. (National Science Foundation)

	Men	Women	
Total doctoral scientists on campus	140,600	26,700	(16%)
tenured faculty	92,700	10,500	(10.2%)
nontenured faculty	19,300	5,600	(22.5%)
nontenure track	12,400	5,300	(29.9%)

that women earned 35 percent of all science bachelor's degrees between 1960 and 1980.

Looking only at scientists with doctorates, 91.7 percent of women Ph.D.'s and 94.9 percent of men Ph.D.'s were in the 1983 labor force. The unemployment rate for women scientists in 1983 was 2.5 percent; for men it was 0.8 percent. (For life scientists, the unemployment rates for women and men were 3.0 and 0.9 percent respectively. The higher rate holds true for all groups of women, no matter what their field or level of experience, no matter whether they are married or single. It is sad but true that though women are about 10 percent of the Ph.D. science labor force, they are also a third of all scientists who are unemployed and looking for work. Interestingly, the unemployment rates for men and women scientists with bachelor's and master's degrees are more similar.

A 1981 National Research Council study reported that more women than men take postdoctoral positions within a year of receiving their Ph.D.'s. They also hold these apprentice-like positions longer because of difficulty in finding suitable faculty or research jobs. Later in their careers, they are slower to gain tenure and more often stuck in lower-level positions. The situation does not seem to have improved in more recent years. By 1983, women accounted for almost 30 percent of all postdoctoral appointments.

Women in Industry

In 1979, more women scientists were employed by universities and colleges than by government and industry. By 1984, the balance had tipped the other way, perhaps largely because of rapid growth in industrial employment. Industry employed most men (64.8 percent) and half the women (50.0 percent) scientists and engineers. In the life sciences, however, the older pattern continued to hold: As Table 12 shows, both men and women were more heavily represented on campus. Table 12 also shows that in the life sciences, men and women are distributed among industry, campus, and government almost identically. Science and engineering as a whole shows a much more marked concentration of women, compared to men, on campus.

A 1980 report of the National Research Council to the Office of Science and Technology Policy, *Women Scientists in Industry and Government*, noted that industry does not treat women as well as men in general. Most women scientists in industry work in basic and applied research; men are twice as likely to be promoted to management positions. Women are paid less, too. And women find it harder to get hired in the first place. Of new doctoral scientists seeking jobs in industry, 79 percent of the men but only 72 percent of the women had jobs on graduation. The sex differences in pay and hiring rates were greatest in the life sciences, which included the greatest number of women doctorates.

Women in Government

The picture is not much better in government. Between 1974 and 1984, women scientists and engineers employed by the federal government increased by 300 percent, from 8,000 to 31,500. Still, women hold a clear minority of the Ph.D.'s in government scientific service. In 1978, women were reaching higher grades and management positions more quickly than men, but only 500 of 17,600 federal managerial jobs were held by women. The disparities in responsibility and pay remain. The pay differential reflects the typically lower rank of women, for the federal government fixes pay by grade or rank. In general, women are hired at lower grade and pay than men and remain, in this sense, inferior to men throughout their careers. It seems clear that the federal government continues to stereotype women and treat them as inferiors, although there are clear signs of improvements.

Table 12. Percent men and women employed by industry, educational institutions and federal government, 1984. (National Science Foundation)

	Science & engineering		Life sciences	
	Men	Women	Men	Women
Industry	64.8%	50.0%	30.8%	29.5%
Educational institutions	12.1%	22.7%	36.1%	40.4%
Federal government	7.9%	6.1%	11.9%	8%

Women in Academia

Like industry and government, universities and colleges also give women less than fully equal treatment. Yet though women may have jobs of lower rank and earn less in academia simply because more women have received their doctorates more recently, similar differences appear even when recently graduated men and women are compared. It seems clear that women do not yet enjoy equality of opportunity or advancement, although many of the obstacles in their way have been removed or diminished. The major obstacle remaining may in fact be Cole's exclusion from informal relationships. Unfortunately, this exclusion cannot easily be corrected by legislation. It must probably remain until men have ceased to think of women as a class apart and begun to treat them not just as equals but as equivalents. This has begun to happen already, for many young men have grown up with some freedom from sexist stereotyping, but it has a long way to go.

What Progress Toward Equality?

Until fairly recently, it seemed that women scientists were making reasonable—if slow—progress in the workplace. This progress showed in the beginning salary offers made to women: In 1974, new women bachelor's graduates in the life sciences were offered 88.4 percent as much as men. By 1981, this had risen to 92.9 percent. Unfortunately, by 1984, women bachelor's graduates with less than one year of experience were averaging $14,000 while men were getting $16,100; that is, they were getting only 88.2 percent of the male average.

The picture is even less rosy for women with older degrees. The salary differentials are larger for women with more years of experience, indicating that once they were locked into a pay scale typical of a less enlightened time, they were stuck with it. It seems, unfortunately, that sexual equality is not retroactive.

The story is similar for Ph.D. graduates. In 1979, women scientists' salaries were significantly below those for men, and the gap widens with years since graduation. But where bachelor's graduates have slipped in their progress toward pay equality, the doctorates have not. Women with Ph.D.'s in the biological sciences and 10 years of experience averaged $4,800 less than men in annual pay in 1980. In 1983, the differential was down to $3,600, showing a $1,200 gain in just three years.

There has been progress. However, discrimination does remain, and women can still expect less reward for the time and effort they put into gaining an education. This is true even in such fields as the behavioral, social, and life sciences, which traditionally have employed most women scientists and which still accept the largest numbers of women.

The progress will undoubtedly continue. It has been uneven in recent years, and the trend has even reversed for some groups, but the cause seems plain: the federal government has cut back on many programs, including National Science Foundation (NSF) efforts to open up scientific careers for women. The federal commitment to equality may not expand again until the 1990's.

However, if you are a woman interested in the biological sciences, you should not let the statistics discourage you. Scientists of either sex enjoy relatively high pay and low unemployment, as well as absorbing, fascinating, fulfilling work. And the latter is really far more important.

Minorities

The term "minorities" is generally taken in two ways: it means both racial and ethnic subgroups of society and any group that is treated as inferior, with or without real reason. In the latter sense, it includes women, who are actually more numerous than men. It also includes the physically handicapped, who have just as much trouble gaining educations and jobs as Blacks or women and deserve it just as little.

Blacks, Hispanics, and Asian and Native Americans together make up 18 percent of the U.S. labor force. However, Blacks and Hispanics are distinctly underrepresented in scientific and technical jobs. Asian Americans, who comprise about two percent of the U.S. labor force, occupy about five percent of the scientific and technical jobs. Native Americans have roughly the same representation in the total labor force and the sci/tech labor force.

These figures conceal some intriguing patterns (see Table 13). Compared to whites, Blacks are heavily concentrated in the social and computer sciences; they also have an edge in mathematics and the physical sciences. Hispanics focus on the social sciences. Native Americans go in more for psychology and the life sciences. Asian Americans are gung-ho for engineering, with an edge in math and the computer and physical sciences.

In addition, careful study of the available statistics reveals that among scientists and engineers as a group, Blacks and Asian Americans split among the industrial, educational, and federal government

Table 13. Field distributions of scientists and engineers by race, 1984 (National Science Foundation).

Field	Whites	Blacks	Hispanics	Asians	Native Americans
Engineering	55%	41%	55%	63%	57%
Social sciences	8	18	12	7	6
Psychology	5	8	5	1	9
Life sciences	9	7	8	6	10
Environ. sci.	3	1	2	1	2
Computer sci.	11	13	10	13	9
Math	2	5	3	3	2
Physical sci.	6	7	5	7	5

employment sectors along almost precisely the same lines as whites. Native Americans are underrepresented in education, and Hispanics are underrepresented in industry. In the life sciences, on the other hand, Hispanics are *over*represented in industry and underrepresented in education, while Native Americans are *over*represented in education and underrepresented in industry. Black life scientists are underrepresented in industry and overrepresented in the federal government. Asian Americans show only slight disparities, exceeding white proportions in education and industry and being exceeded by whites in federal employment.

Minority employment in the sciences improved considerably between 1973 and 1983. The total number of doctoral scientists went from 184,000 to 307,800, up 67 percent, with whites rising by 64 percent. Black scientists increased 137 percent, Asian Americans 164 percent, Hispanics 214 percent, and Native Americans 300 percent.

However, in the life sciences, the picture is slightly less rosy. Minority employment increasd there, too, but not by quite so much, as Table 14 shows. Native Americans, particularly, did not fare well in the life sciences. Blacks did especially well in the medical sciences.

Through the 1970s, Black medical school enrollment climbed from 2.8 to 9.1 percent, which is close to the Black proportion in the U.S. population and indicates considerable success in overcoming past discrimination. Unfortunately, this success may not long continue. Black enrollments in colleges and universities are declining more rapidly than enrollments of other groups (see Table 15).

Table 14. Increases in minority employment in the life sciences, 1973-1983 (National Science Foundation).

Field	Whites	Blacks	Hispanics	Asians	Native Americans
Life sciences	59%	83%	117%	162%	0%
Bio. sciences	46	17	543	147	**
Ag. sciences	52	*	200	167	**
Med. sciences	109	300	200	183	**

*1973 numbers too few to estimate.
**1973 and 1983 numbers too few to estimate.

Table 15. Total enrollment (in thousands) in institutions of higher education by race/ethnicity of student, fall 1980 vs. fall 1984 (Source: U.S. Department of Higher Education, Center for Statistics).

	1980	1984	Percent
All institutions	12,087	12,162	0.6%
White	9,833 (81.3%)	9,767 (80.3%)	-0.6
Black	1,107 (9.19)	1,070 (8.8)	-3.3
Hispanic	472 (3.9)	529 (4.3)	12.1
Asian	286 (2.4)	382 (3.1)	33.6
American Indian	84 (0.7)	83 (0.7)	-1.2
Nonresident alien	305 (2.5)	332 (2.7)	8.9

Minority women have also made considerable gains. They are now about 12 percent of all employed women scientists and engineers, or about one percent of all employed scientists and engineers. Among these women, Blacks are better represented than among men; Black women also account for a larger share of Black doctorates than do the women of other groups in those groups. Field distributions for the women are roughly similar to those for their groups as wholes (see Table 13) with the notable exception that a third of all Native American women scientists go into the life sciences.

What will happen to minority women in the future? It seems likely that their lot will continue to improve in the sciences. Among Blacks, the campus enrollments of women are declining much less than those of men. Among other minority groups, the women are increasing their presence on campus much more rapidly than men. And these differences cannot help but show up in the scientific and technical workplace a few years hence.

Past and future progress in the fight against discrimination depends on active efforts ("affirmative action") to find, train, and recruit minority members. The National Institutes of Health (NIH) has been active in this effort, with its Minority Access to Research Careers (aimed at improving minority faculty members' research skills) and Minority Biomedical Support (aiding students and faculty) programs. The AAAS OOS has promoted conferences on the problems and opportunities facing minority scientists. With grants from the Ford Foundation and the U.S. Department of Education's Minority Institutions Science Improvement Program, it began in 1981 a program called Mathematics, Engineering, Science, and Health Network of Minority Professional Associations (MESH-work) to recruit minority students (see the OOS's newsletter, *MESH-work News*, for reports of new data, policies, organizations, and professional opportunities). The OOS also works on behalf of the handicapped and women "to accelerate the progress of groups underrepresented in science."

Opportunities do exist in science and in biology for people of every kind. No group has a monopoly on talent, and most biologists know this. Yet underrepresentation is real. Where does the problem lie? Is it discrimination? To some extent, yes. But discrimination *has* been largely done away with. The problem today is much more one of a shortage of recruits, of students who believe there are places in science for them. Many "affirmative action" programs strive to solve this problem, but they do face a basic obstacle. The image of science is white and male, and young minority students rarely learn of successful scientists who belong to their particular minority.

They may know of George Washington Carver (the peanut man), but if they have ever heard of Percy Julian, who developed a synthetic drug for glaucoma, they have not heard that he too was black. Too many minority scientists lose their minority labels when they gain recognition for their work. No one claims them as whites, but they do become "just another human being." And while this may mesh well with the ideal of every fair-minded person, it does deny the young minority student a role-model and a message of potential success.

Perhaps we should be sure to recognize and advertise the minority affiliations of scientific success. This would encourage the young, though some people would surely call it demeaning or condescending. They might prefer to trust the statistics: every year sees more minority scientists, more college-educated minority parents, and a larger proportion of minority youth growing up in a setting that encourages intellectual aspiration. And the statistics *can* be trusted. If nothing happens to set progress back, equal opportunity should in time, and with no further help, become a reality. However, if we can possibly speed that day by being free with our encouragement and support, do we not have a duty to do so? Many people say we should, and this is in fact one function of the OOS. It is also a function of the many professional associations dedicated to the interests of minority biologists (listed at the end of this chapter). It has also been a function of many federal programs, but the Reagan administration claims that minority-directed science education efforts are ineffective. NSF's Science and Engineering Education and Equal Opportunities in Science and Technology programs, the Department of Education's Women's Educational Equity Act Program, the Environmental Protection Agency's minority institutions research program, and NIH's Minority Access to Reseach Careers Program have all been threatened, hamstrung, or killed. Only NIH's Minority Biomedical Support Program seems to be holding its own, for the time being.

What needs to be done? We have already mentioned the OOS survey of minority women scientists. The OOS report of that survey, by Paula Quick Hall, concluded that we should adopt three major educational objectives:

1. Give all students the most rigorous possible math and science education.
2. Give all students enough information to make wise educational and career decisions.
3. Motivate students to persist until they live up to their intellectual potentials.

To achieve these objectives, the report adds, we must:

1. Improve access to career information
2. Increase interaction of students with appropriate role-models
3. Emphasize self-discipline
4. Increase extracurricular science educational activities
5. Introduce science and math earlier in the schools
6. Emphasize more "hands-on" science
7. Use more minority and women scientists as speakers
8. Use more educational materials pertinent to the students' experience
9. Support more supplementary academic programs
10. Improve career guidance and counseling
11. Give college students more study-related work
12. Involve students more often in research
13. Make more appropriate adademic and personal support services available
14. Recruit more women and minorities as science teachers
15. Sensitize faculty to the problems of minority students at majority institutions
16. Increase faculty awareness of the potential of minority students

We have made some progress, but not so much that we should not continue our efforts to help women and minorities. I therefore wish to end this chapter with a listing of career and counseling information for these people, taken from the Bureau of Labor Statistics' *Occupational Outlook Handbook*, 1986-1987 edition. The list begins with the *Directory of Counseling Services*, published by the American Association for Counseling and Development, 5999 Stevenson Avenue, Alexandria, VA 23304; it is available in many libraries and guidance offices.

The physically handicapped can write or call the President's Committee on Employment of the Handicapped, 1111 20th Street, NW, Room 636, Washington, DC 20036 (202-653-5044). The blind and deaf-blind can obtain toll-free job information by calling 1-800-638-7518.

Minorities can contact the League of United Latin American Citizens, National Educational Service Centers, 400 First Street, NW, Suite 176, Washington, DC 20001 (202-347-1652), or the National Association for the Advancement of Colored People (NAACP), 186 Remsen Street, Brooklyn, NY 11201 (718-858-0800). Blacks especially will also find the magazine *The Black Collegian* helpful (write or call Black Collegiate Services, Inc., 1240 South Broad Street, New Orleans, LA 70125; 504-821-5694).

Women can write or call the U.S. Department of Labor, Women's Bureau, 200 Constitution Avenue, NW, Washington, DC 20210 (202-523-6652). Another place to find information is Catalyst, 250 Park Avenue South, New York, NY 10003 (212-777-8900). Women can also contact Wider Opportunities for Women, 1325 G. Street, NW, Lower Level, Washington, DC 20005 (202-638-3143).

Associations for Minority Biologists

National Dental Association, Suite 24, 5506 Connecticut Ave.,
Washington, D.C. 20015

Promotes the interests of minority dentists.

Association of American Indian Physicians, Suite 206, 6801
S. Western, Oklahoma City, OK 73139

Works to increase the number of American Indian health
professionals in the Indian health delivery system.

National Black Nurses Association, Inc., P.O. Box 18358,
Boston, MA 02118

Fosters nursing care for Blacks and serves as a professional
organization for Black nurses.

National Medical Association, Suite 310, 1301
Pennsylvania Ave., Washington, D.C. 20004

Strives to eliminate religious and racial discrimination and
segregation from American medical institutions.

National Society of Allied Health, 300 W. 43 St., New York,
NY 10036

Fosters education in the health professions for minority
students.

Student National Medical Association, Suite 1000, 1133 15 St.,
NW, Washington, D.C. 20005

Works to increase the number of minority physicians.

National Pharmaceutical Association, Howard University College
of Pharmacy, 2300 4 St., NW, Washington, D.C. 20059

Unites minority pharmacists.

African Scientific Institute, P.O. Box 29119, Washington, D.C.
20017

Promotes recruitment and professional development of Black
scientists and engineers.

Associations for Minority Biologists (continued)

American Indian Science and Engineering Society, 35 Porter Ave., Naugatuck, CT 06770

Works to increase the numbers of and increase opportunities for American Indian scientists and engineers.

National Institute of Science, Department of Chemistry, Central State University, Wilberforce, OH 45384

Provides a means for the exchange of scientific information and the improvement of science education.

National Network of Minority Women in Science, Office of Opportunities in Science, AAAS, 1776 Massachusetts Ave., NW, Washington, D.C. 20036

Supports the education and advancement of minority women in science.

Organization of Black Scientists, Inc., Department of Chemistry, Atlanta University, 223 Chestnut St., SW, Atlanta, GA 30314

Identifies, provides, and administers the research and educational goals of Black scientists.

Society for Advancement of Chicanos and Native Americans in Science, P.O. Box 30040, Bethesda, MD 20814

Strives to increase the participation and recognition of Chicanos and Native Americans in science.

Career Pamphlets

Ask Any Forester

Society of American Foresters
5400 Grosvenor Lane
Bethesda, MD 20814

*Career Development
Opportunities for
Native Americans*

Bureau of Indian Affairs
U.S. Department of the Interior
1951 Constitution Ave., NW
Washington, D.C. 20245

*Career Opportunities
for the Ichthyologist*

American Society of Ichthyologists
and Herpetologists, Inc.
NMFS Systematics Laboratory
National Museum of Natural
History
Washington, D.C. 20560

*Career Opportunities
in Ornithology*

American Ornithologists' Union
National Museum of Natural
History
Washington, D.C. 20560

*Careers in
Biochemistry*

American Society of Biological
Chemists
9650 Rockville Pike
Bethesda, MD 20014

*Careers in Biological
Systematics*

Society of Systematic Zoology
c/o Department of Entomology
National Museum of Natural
History
Washington, D.C. 20560

Careers in Biology

Education Department
American Institute of Biological
Sciences
1401 Wilson Blvd.
Arlington, VA 22209

Careers in Physiology	American Physiological Society 9650 Rockville Pike Bethesda, MD 20014
Careers in the Poultry *Industry: A Job Is* *Ready When You Are*	National Broiler Council 1155 15th St., NW, Suite 614 Washington, D.C. 20005
Door to the Future: *Careers for Women*	U.S. Department of the Interior Central Employment Office, Room 2640 18th & C Streets, NW Washington, D.C. 20240
A Guide to Careers in *Science Writing*	National Association of Science Writers, Inc. P.O. Box 294 Greenlawn, NY 11740
Careers in Physical *Therapy*	American Physical Therapy Association 1156 15th Street, N.W. Washington, D.C. 20005
The Modern Anatomist	American Association of Anatomists c/o Dr. William P. Jollie MCV Station, Box 101 Richmond, VA 23298
Dentistry—A *Changing Profession*	Council on Dental Education American Dental Association 211 East Chicago Avenue Chicago, IL 60611
Plant Pathology: A *Scientific Career* *for You*	American Phytopathological Society 3340 Pilot Knob Road St. Paul, MN 55121

Microbiology in *Your Future*	American Society for Microbiology 1913 I Street, N.W. Washington, D.C. 20006
Careers in Botany	Dr. Patricia K. Holmgren Secretary Botanical Society of America The New York Botanical Garden Bronx, NY 10458
Careers in Animal *Biology*	American Society of Zoologists Box 2739 California Lutheran College Thousand Oaks, CA 91360
Dieticians: The *Professionals in* *Nutritional Care*	The American Dietetic Association 430 North Michigan Avenue Chicago, IL 60611
Careers in Plant *Physiology*	American Society of Plant Physiologists 9650 Rockville Pike Bethesda, MD 20814
Careers in Psychology	American Psychological Association 1200 Seventeenth Street, N.W. Washington, D.C. 20036
Careers in *Demography*	The Population Association of America, Inc. P.O. Box 14182 Benjamin Franklin Station Washington, D.C. 20034
Helping Hands: *Horizons Unlimited* *in Medicine*	American Medical Association 535 N. Dearborn St. Chicago, IL 60610

Nutrition: The Challenge to Improve the Quality of Life	American Institute of Nutrition 9650 Rockville Pike Bethesda, MD 20814
Pathology as a Career in Medicine	Intersociety Committee on Pathology Information 4733 Bethesda Ave. Bethesda, MD 20814
The Science of Mammalogy	American Society of Mammalogists National Museum of Natural History Washington, D.C. 20560
This Is the Profession of Pharmacology	American Society for Pharmacology and Experimental Therapeutics, Inc. 9650 Rockville Pike Bethesda, MD 20014
200 Ways to Put Your Talent to Work in the Health Field	National Health Council, Inc. 70 W. 40th St. New York, NY 10018
A World of Opportunity— Programs of the Public Health Service	Public Health Service Personnel Staffing 5600 Fishers Lane, Room 14-A40 Rockville, MD 20857

You may obtain any of the above pamphlets by writing to the listed addresses. If you want an even broader choice of pamphlets on biological careers, write to Joy Gold, Department of Vertebrate Zoology, National Museum of Natural History, Smithsonian Institution, Washington, D.C. 20560, and ask for the bibliography "Sources for information on careers in biology, conservation, and oceanography." You can obtain a list that offers less on biological careers but more on other careers in science by writing for "Career information sources" to the American Association for the Advancement of Science, 1776 Massachusetts Avenue, NW, Washington, D.C. 20036.

Further Readings

Akin, J.N., *Teacher Supply and Demand 1981 through 1985,*
Association for School, College and University Staffing, 1981-
1985.

Babco, E.L., *Salaries of Scientists, Engineers, and Technicians: A
Summary of Salary Surveys* (Washington, D.C.: Scientific
Manpower Commission, November 1985). (The Scientific
Manpower Commission is now the Commission on
Professionals in Science and Technology.)

Boesch, M.J., *Careers in the Outdoors* (New York: Dutton, 1975).

Broad, W.J., "Slave labor on campus: The unpaid postdoc," *Science,*
216 (May 14, 1982), 714-715.

Campbell, P.N., ed., *Biology in Profile* (New York: Pergamon,
1981).

Carter, C., *Women in the Sciences* (Bibliography), L.C. Science
Tracer Bullet, Library of Congress, Washington, D.C., 1976.

Cole, J.R., "Women in science," *American Scientist,* 69 (July-
August 1981), 385-391.

Cole, J.R., and H. Zuckerman, "Marriage, motherhood and research
performance in science," *Scientific American,* February 1987,
119-125.

*Confronting AIDS: Directions for public health, health care, and
research* (Washington, D.C.: National Academy Press, 1986).

Culliton, B.J., "The academic-industrial complex," *Science,* 216
(May 28, 1982), 960-962.

Easton, T.A., and R.W. Conant, *Cutting Loose: Making the
Transition from Employee to Entrepreneur* (Chicago: Probus,
1985).

Easton, T.A., and R.W. Conant, *Using Consultants: A Consumer's
Guide for Managers* (Chicago: Probus, 1985).

Fanning, O., *Opportunities in Environmental Careers* (Skokie,
Illinois: National Textbook Co., 1981).

Goldin, S., and K. Sky, *The Business of Being a Writer* (New York:
Harper & Row, 1982).

Hall, P.Q., *Problems and Solutions in the Education, Employment
and Personal Choices of Minority Women in Science,* Office of
Opportunities in Science, AAAS, August 1981.

Hawkins, D.E., and P.J. Verhaven, *Utilization of Disadvantaged Workers in Public Park and Recreation Services*, American Association for Health, Physical Education and Recreation and National Recreation and Park Association, 1974.

Hodge-Jones, J., ed., *Minority Student Opportunities in United States Medical Schools 1980-1981*, 6th ed., Office of Minority Affairs, Division of Student Programs, Association of American Medical Colleges, Washington, D.C., 1980.

Kidd, C.V., "New academic positions: The outlook in Europe and North America," *Science*, 212 (April 17, 1981), 293-298.

Lewis, R., "Medical detectives" (CDC), *Biology Digest*, 8 (March 1982), 10-22.

Medawar, P.B., *Advice to a Young Scientist* (New York: Harper & Row, 1979).

National Science Foundation, *Women and Minorities in Science and Engineering*, Report NSF 86-301, January 1986.

Occupational Outlook Handbook, 1986-1987 Edition, Bureau of Labor Statistics, U.S. Department of Labor (Washington, D.C.: U.S. Government Printing Office, 1986).

Postdoctoral Appointments and Disappointments, National Research Council, Washington, D.C. 1981.

Puerto Ricans in Science and Biomedicine, Office of Opportunities in Science, AAAS, Washington, D.C., 1981.

"Surgeon General's report on Acquired Immune Deficiency Syndrome," available free of charge from AIDS, P.O. Box 14252, Washington, D.C. 20044.

Vetter, B.M., *A Statistical Report on Black Americans in Science*, Office of Opportunities in Science, AAAS, Washington, D.C., 1981.

Vetter B.M., *The Technological Marketplace: Supply and Demand for Scientists and Engineers* (Washington, D.C.: Scientific Manpower Commission, May 1985).

Vetter, B.M., "Women scientists and engineers: Trends in participation," *Science*, 214 (December 18, 1981), 1313-1321.

Biographies

Benison, S., A.C. Barger and E.L. Wolfe, *Walter B. Cannon: The Life and Times of a Young Scientist* (Cambridge, Massachusetts: Belknap/Harvard University Press, 1987).

Chargaff, E., *Heraclitean Fire: Sketches from a Life Before Nature* (New York: Rockefeller University Press, 1978).

Clark, R.W., *The Life of Ernst Chain: Penicillin and Beyond* (New York: St. Martin's, 1985).

Dubos, R., *The Professor, the Institute, and DNA* (New York: Rockefeller University Press, 1976).

Eccles, J.C. and W.C. Gibson, *Sherrington: His Life and Thought* (New York: Springer-Verlag, 1979).

Haas, R.B., *Muybridge: Man in Motion* (Berkeley and Los Angeles: University of California Press, 1976).

Kundsin, R.B., ed., *Women and Success: The Anatomy of Achievement* (New York: Morrow, 1974).

Nisbett, A., *Konrad Lorenz* (New York: Harcourt Brace Jovanovich, 1976).

Ogilvie, M.B., *Women in Science: Antiquity through the Nineteenth Century* (Cambridge, Massachusetts: MIT Press, 1986).

Selye, H., *The Stress of My Life: A Scientist's Memoirs*, 2nd ed. (New York: Van Nostrand Reinhold, 1979).

Shine, I. and S. Wrobel, *Thomas Hunt Morgan: Pioneer of Genetics* (Lexington, Kentucky: University Press of Kentucky, 1976).